AUSTRALIAN Grasses

A Guide to Native Grasses, Sedges, Rushes and Grasstrees

SECOND EDITION

Nick Romanowski

PUBLISHING

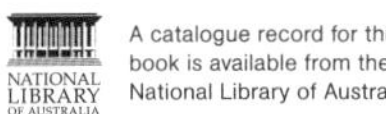

ISBN: 9781486320523 (pbk)
ISBN: 9781486320530 (epdf)
ISBN: 9781486320547 (epub)

How to cite:

Romanowski N (2026) *Australian Grasses: A Guide to Native Grasses, Sedges, Rushes and Grasstrees*. 2nd edn. CSIRO Publishing, Melbourne.

Published by:

CSIRO Publishing
36 Gardiner Road, Clayton VIC 3168
Private Bag 10, Clayton South VIC 3169
Australia

Telephone: [+613] 9545 8555
Email: csiropublishing@csiro.au
Website: www.publishing.csiro.au
Sign up to our email alerts: publishing.csiro.au/earlyalert

Front cover: *Phragmites australis* (photo by Nick Romanowski)
Back cover: (left to right) *Cenchrus purpurascens, Juncus ingens, Cyperus bifax* (photos by Nick Romanowski)

Edited by Adrienne de Kretser, Righting Writing
Cover design by Cath Pirret
Typeset by Envisage Information Technology
Printed in China by 1010 Printing International Ltd

CSIRO acknowledges the Traditional Owners of the lands that we live and work on across Australia and pays its respect to Elders past and present. CSIRO recognises that Aboriginal and Torres Strait Islander peoples have made and will continue to make extraordinary contributions to all aspects of Australian life including culture, economy and science. The use of Western science in this publication should not be interpreted as diminishing the knowledge of plants, animals and environment from Indigenous ecological knowledge systems.

The paper this book is printed on is in accordance with the standards of the Forest Stewardship Council® and other controlled material. The FSC® promotes environmentally responsible, socially beneficial and economically viable management of the world's forests.

Nov25_01

Contents

Acknowledgements

WHEN I WROTE MY FIRST book on grasses it was a struggle to find native and indigenous gardens worth photographing, and it was difficult to separate what was known about native grasses in particular, from some of the misinformation that was widely believed. For this book, I have been fortunate in being able to draw upon a wealth of information sources, diverse contacts across many parts of Australia and a wide choice of gardens using native grasses, sedges and related plants.

I would particularly like to acknowledge the late Surrey Jacobs who long acted as a courteous personal mentor, steering me past the assumptions and black holes in my knowledge which someone whose primary background is in zoology can easily fall into. Karen Wilson at the New South Wales Herbarium has identified many sedges over the years, including introduced weeds still being sold under new 'indigenous' names. I am also grateful to the many staff at the Melbourne Herbarium, particularly Val Stajsic and Neville Walsh, for their help in identifying species over the years.

Botanic gardens are leading the way towards a more indigenous style of gardening, and this book would be very much poorer without the work done by people like Brady Grand in Kings Park, Perth, who also identified a number of photos I took around south-west Australia. In my local corner of Australia (the Otway Ranges, south-west of Melbourne), I would particularly like to acknowledge John Arnott in his role as curator of Geelong Botanic Gardens, where native grasses and their relatives were prominently (and imaginatively) displayed in a range of ways, from smoothly integrated in and among long-established exotic plantings, to semi-formal beds to the Japanese garden shown in Chapter 2.

For this edition, I offer particular thanks to Anna Povey who has corrected my explanation of the structure of grass flowers, which was clumsily described in the original edition. Andrew Orme at the New South Wales Herbarium was helpful with information on *Dracophyllum* and its relatives on mainland Australia. Lyle Filippe of Roraima Nursery and his co-workers create all manner of magical animal sculptures, including the emu that graces the wall of the author's swimming dam (see Chapter 2).

Thanks also to Nola and Michael Fenech at Wallis Creek Water Gardens, Durnford Dart at Bamboo Australia, Leanne Westen at Kuranga Native Nursery, and many staff at Fairhills Nursery in Queensland for very useful discussions about tropical grasses.

Some of the gardens described in this book remain unattributed because no-one was home at the time I passed by, meaning that I was unable to talk to the owners about their inspirations and ideas then and there, and didn't receive a phone call in response to the note I left. I'd be pleased to hear from anyone I've not yet been able to acknowledge, and would include those attributions in future editions of this book. I'd also like to add more photos of new gardens using native grasses!

Why native grasses?

AT SOME TIME IN THE late 1980s grasses took the gardening world by storm, and for the next decade almost every Australian nursery seemed to be importing and propagating every exotic species they could lay their hands on while native grasses, sedges, rushes and their relatives remained virtually unknown. The only book during this time that covered any significant number of native species was my own *Grasses, Bamboos and Related Plants in Australia* (1993), although it also included the many exotics being sold in this country.

Even back then there were ominous signs that the fashion for exotics set by gardeners in the northern hemisphere was inappropriate for Australian conditions. I warned that many of the plants on offer were already established as environmental, agricultural or garden weeds in at least some of the diverse climates of this large country, while many others had the potential to become problematic.

Thirty years later, native species are increasingly appreciated as low-maintenance lawns which remain green when exotic lawn grasses have shrivelled and dried. They are being replanted in pasture lands as drought-tolerant grazing, they are being used for a wide range of purposes from spectacular, tough, long-lived garden specimens to low maintenance ground covers, and some are even being considered as potential perennial food crops. And many native grasses are as spectacularly beautiful as the finest grasses the rest of the world has to offer.

This book brings together decades of learning about diverse aspects of the Australian grassy flora, illustrated with photos I have taken in rainforests, wetlands and mallee scrub, from mountains to the coast and arid inland areas, in botanic gardens and water gardens from Tasmania to the tropics. Well over 200 species and forms are described in the encyclopedia which makes up much of this book. Many of these are already available or are familiar from wild places, and their potential uses and ecological values beyond the garden will only be touched upon in passing as they would warrant a hefty book in their own right.

This second and much expanded edition of *Australian Grasses* will offer some very different ways of looking at gardens and gardening. If you're concerned about what introduced plants and attitudes to gardening have done to Australia, about our excessive use of water and dependence on fertilisers which have well-documented detrimental effects on wetlands and waterways, or about the ever-increasing array of introduced weeds which do nothing for the environment, agriculture or gardens – this book will help you create a garden that is far better suited to Australian conditions.

The elegant stems of Slender bamboo-grass (*Austrostipa verticillata*)

1

What are grasses, sedges and rushes?

GRASSES, SEDGES, RUSHES AND CORDRUSHES, along with a few smaller plant families, are a natural and fairly closely related group of plants which most people would intuitively group together as 'grasses'. Some of their more distant relatives in other families such as various lilies, cumbungis, water ribbons and others which share obviously grass-like features are also included (collectively, these are monocotyledons, plants which only produce a single leaf when their first seed germinates), but the overall emphasis in this book is on the true grasses and their close relatives, with their spare and simple beauty.

These aren't plants that draw your attention with gaudy flowers, nor are they ancient species worthy of veneration. Their history in geological time is relatively short and is often closely tied to the evolution of mammals, of which we are the latest and most destructive type. Most grasses have evolved to cope with the chewing teeth of grazing animals, and are able to regrow from the stubs left behind by animals which evolved only because there were grasses for them to browse upon. To be fair, mammalian teeth are worn away very rapidly by the silicates in grasses, so the relationship between plants and their predators has been more of a battlefield than anything like co-operation!

Many grasses are opportunists – quick to move into places where taller, shading vegetation has been removed or thinned by fire, lava flows or the other disruptions of the natural world which have fuelled change over the past millions of years. They are mostly pollinated by wind, or will self-pollinate when no other opportunity arises, and this keeps the many species alive through harsh times. As very few of these plants need insects to spread their pollen, they don't have colourful flowers and rarely are fragrant or otherwise attractive to insects and other potential pollinators.

Most groups of grassy plants grow in one of two basic ways. They either form tussocks that stay tidily put in one place and gradually reach a predictable maximum size, or a single plant may spread more widely (run) to cover anything from a small and well-defined area to many hectares, depending on the species. Not all grasses have adapted to grazing. The more primitive species include bamboos, whose tall canes carry the leaves up into the forest canopy where there is more light.

Gardeners and bushwalkers are often attracted by the movement of grass flowerheads blowing in the wind, or the utter stillness of those very same heads on a quiet autumn day. Some grasses and sedges have wide-open flowerheads, others have them packed into tidy wands or loose groups of thousands, or in clusters or plumes or any other term you can think of to try to describe their great diversity.

This is not a botanical treatise, so I will use several informal terms which should make perfect sense to anyone who wants to understand these beautiful plants. The next section describes terms with more technical precision. A **flowerhead** is a young flowering stalk, around the time when it is ready for the newly opened flowers to be pollinated. A **seedhead** is the later stage when the seeds are ripening, though from a distance these two stages may not look very different in some species. **Flowerstems** are the stalks that hold both flowerheads and (later) the seedheads aloft.

It is also worth defining several terms which are used throughout the text in various ways. The term **native** is a loose category, referring to any plant which was already growing in Australia before European settlement in 1788. This covers a multitude of species across the continent, but it's important to remember that although a grass from Western Australia may be native there this doesn't mean it is native in eastern Australia as well. Plants which are native to your local area are **indigenous**, truly native in every sense of the word, though they may also be much more wide-ranging. Finally, **exotic** means much the same thing as **introduced** – in most cases brought in from other countries, though these terms also apply to any native grass which has been moved outside its natural range.

GRASSES – FAMILY POACEAE

Grasses generally have hollow, cylindrical stems, with the leaves attached alternately at the nodes by a stem-hugging sheath. **Nodes** are joints where there is a solid partition to give more strength to the flexible stem. These are particularly obvious in the largest grasses, the bamboos. Botanists use the descriptive term **floret** for each individual flower in a grass flowerhead, and each floret will produce just one seed within a pair of papery, protective bracts known as a palea and a lemma. A **spikelet** is a group of florets (sometimes there may only be one) found along a shared stem. These are arranged on opposite sides, so grass spikelets are generally distinctly flattened and bi-symmetrical. And at the base of every spikelet there is an extra pair of papery bracts with no florets inside them; these are the upper and lower glumes.

Grass spikelets: *Glyceria australis* above, the awned *Amphibromus nervosus* below

The papery sheaths protecting each floret and later its developing seed may end with a long, slender awn projecting from the tip, which can most easily be seen in ears of wheat or barley. This can also be useful in identification; for example, native *Glyceria* and *Amphibromus* grow in similar habitats and often look very similar from a distance, but the florets of *Amphibromus* have a prominent awn when you look more closely.

SEDGES – FAMILY CYPERACEAE

Sedges are very closely related to grasses, though they are more likely to have cylindrical spikelets than the flattened, bi-symmetrical type seen in grasses. The papery bracts of sedges are called **glumes**. Each one usually encloses a single flower, and the glumes are often arranged spirally around the spikelet. Sedge flowerheads may be as simple as a single spikelet at the tip of a stem (see *Eleocharis*), or they may be substantial, showy masses of spikelets held high on a strong flowerstem. These stems are often (but not always) triangular, and unlike those of grasses they are usually solid. Many sedges have reduced leaves or none at all, so their stems may resemble some of the more upright rushes though their flower structure is very different (see the next section). Although many sedges

River clubrush (*Schoenoplectus tabernaemontani*) spikelets

What's in a name?

The common names of sedges, rushes and reeds and their various parts are applied apparently at random. Understandably, this can cause confusion. All plants in the family Cyperaceae are technically sedges, while the true rushes are family Juncaeae and Common reed is a grass. However, these names are loosely applied in many other ways – clubrushes, twigrushes and leafrushes are all sedges, while bulrushes are the very distinctive family Typhaceae and burr-reeds are yet another family called Sparganiaceae. This is why this book emphasises the biological differences between the various plant families. Although they may look confusing at first glance, there are only a few families involved, and their botanical names are a guide to their true, natural relationships.

are well adapted to dry conditions, most of them are associated with water. This family includes a great number of the ecologically important grassy plants of wetlands.

RUSHES – FAMILY JUNCACEAE

Rushes can be almost as variable in size and shape as the sedges and grasses but they have a very different type of flower structure. The small flowers are often grouped into clustered heads and are surrounded by six tepals so they look like a tiny, not very colourful, six-petalled flower. Once fertilised, the central part of each flower becomes a capsule which splits open along several seams, to shed many *very* small seeds as it opens. The shape of the mature capsules, and whether they are tucked inside the tepals or protrude beyond them, are useful aids to identification in this notoriously difficult group. Many of the larger rushes of interest to gardeners are more-or-less leafless. Useful identification features include some clinging, papery sheaths at their base, along with the number of fine ridges along the stems and the texture of the inside of the stems. The stems themselves are usually hollow, with a foamy filler that helps to strengthen them while still allowing considerable movement.

Pale rush (*Juncus pallidus*) seed capsules developing

CORDRUSHES – FAMILY RESTIONACEAE

Cordrushes are a southern hemisphere family, with two areas of greatest diversity in southern Africa and southern Australia. Native species are particularly abundant and ecologically important in Western Australia, growing in a variety of habitats from dry scrub to the fringes of wetlands. There are also quite a few species in eastern Australia. The 'stems' of this family are technically branches which spring from an underground stem;

Tassel cordrush (*Baloskion tetraphyllum* subsp. *tetraphyllum*), male flowerheads above, female flowerheads below

they may be flat or cylindrical, solid or hollow. They are often clearly marked at the nodes with dark sheaths which are all that remains of the leaves, though there may be true leaves higher up as well.

Nearly all species of cordrushes are male *or* female, which means you need both sexes growing together if you plan to collect seed for propagation. As male and female spikelets are often very different in appearance, they may look like different species at first glance. Despite their great abundance in a range of habitats, most cordrushes are so little known that they have no common names.

GRASSTREES (FAMILY XANTHORRHOEACEAE) AND MAT-RUSHES

Although it isn't obvious at first glance, grasstrees (*Xanthorrhoea* and *Kingia*) and mat-rushes (*Lomandra*) are fairly closely related – in a sense the grasstrees could be described as specialised mat-rushes with trunks. Not all that long ago they were grouped together in the Xanthorrhoeaceae, but a more recent system turned everything but *Xanthorrhoea* out of this family, shifting *Lomandra* and *Kingia* to the Dasypogonaceae. *Lomandra* has also been regarded as a relative of

Yellow-faced honeyeater feeding from a flowering Grasstree (*Xanthorrhoea australis*)

asparagus! With these families we are moving away from true grasses and their relatives and coming perilously close to lilies and their diverse kin, as can be seen if you look closely at a grasstree when its star-like flowers are first opening.

OTHER GRASS-LIKE PLANTS

From a gardener's viewpoint some smaller but closely related families are also very grass-like. For example, whip vine (*Flagellaria*), burr-reeds (*Sparganium*) and cumbungis (*Typha*) are close relatives, and some more distant relatives from the families Philydraceae, Juncaginaceae, Xyridaceae and even many lilies and their relatives have grass-like foliage. Many of these produce attractive flowers for a short season as an added bonus. A selection of these distant relatives has been included in this book for completeness, along with a few real outliers such as Giant grasstree (*Richea pandanifolia*, the world's largest heath) and its smaller relatives, along with their tropical namesakes the screwpines (*Pandanus*).

2

Gardening with grasses: grasslands, lawns and formal gardens

Natural garden style: *Gahnia sieberiana* with emu nest

AUSTRALIAN GRASSES AND THEIR DIVERSE relatives have been used in many ways, from the stylised, low-maintenance simplicity of drought-tolerant lawns that won't invade nearby bushland, to interplanting among native flowering plants to give an impression of true, diverse wilderness. They have been used to soften the edges of gardens of introduced plants, to bring new vitality to the tired cliches of the so-called nature-strip and to recreate the floral wealth of native grasslands. There are diverse non-weedy indigenous grassy species which make fine replacements for nearly all the introduced grassy weeds which were fashionable not all that long ago, others which are even more attractive, and still others which could ultimately become the basis of a new style of gardening.

Many native grasses and related plants will flourish as dramatic yet tidy formal specimens in pots and containers, while others fit well into the studied naturalism of a Japanese-style garden – and nearly all of them are more tolerant of drought, heat and neglect than their relatives from other countries. At yet another extreme, gardeners may enjoy emphasising their rougher textures and qualities – native grasses are increasingly used to soften the boundaries between the wilderness and the orderly gardens of some very expensive resorts, bridging the gap between the 'needs' of the world of tourism and the surrounding wilderness.

Formal garden style: *Cyperus vaginatus* with Japanese cranes in Geelong Botanic Gardens, Victoria

The potential uses of grasses in the garden are legion. Gardeners should choose according to their personal taste rather than follow whatever is the current fashion, as the latter practice tends to create uninspiring monocultures of just a few, all-too-familiar species. Wherever we look in the natural world there are diverse grasses, and most gardens will only benefit from a greater diversity of species. Many of the uses and potentials of these plants are sketched out in this chapter, and the specific needs and preferences of the examples are discussed in more detail in the encyclopedia which makes up much of this book.

Some of the plants described in the encyclopedia can also be used as cut flowers. The flowerstems of plants as different as *Baloskion*, *Gahnia* and the ubiquitous cumbungis last well in water if they are picked at the right time and can be dried out to use in more permanent arrangements. Many species have been used for basketry and other forms of weaving; once again, cumbungis star here and they have even been used as a reasonably long-lasting upholstery or mesh for chairs. Basket sedge (*Carex tereticaulis*) was a favoured source of materials for long-lasting baskets and fish traps among Aboriginal peoples of southern Australia.

NATURAL STYLE – LEARNING FROM GRASSLANDS

Before looking at some of the more familiar ways in which grasses are used in gardens we should consider their most fundamental yet unfamiliar and neglected aspect: grasses as the defining plants of native grasslands. These once covered huge areas of Australia but have been reduced to ~1% of their original range today, so most Australians are only dimly aware of their former glories. Yet these species-rich domains can be as gaudy in the flowering season as some of our better-known national parks, and offer some very different ideas and perspectives on ways to approach gardening with indigenous plants.

When European settlers first reached the drier inland areas of Australia, they were impressed to find extensive and succulent grasslands in areas where there was not enough rainfall to support trees. Even among the taller forests there were often dense stands of grasses beneath the canopy. Deliberate burning by Aboriginal peoples had helped to maintain a mosaic of forested and grassy areas that attracted a diversity of grazing animals such as kangaroos and wallabies, and probably made it easier to hunt them as well. The new arrivals were primarily interested in grazing for livestock. Native grasslands seemed excellent for this purpose, but within a few years sheep had grazed and compacted many of

Speargrasses, mostly *Austrostipa mollis*

the indigenous food plants into virtual extinction. The newcomers learnt the limits of Australia's carrying capacity the hard way.

Over time, native grasslands were gradually replaced with introduced grasses which were supposedly more palatable or more productive. However, in most cases the alien pastures turned out to be not well adapted to Australian conditions, alternating between winter lushness and summer drought. Without exception, the introduced species which most effectively competed with indigenous grasses in difficult conditions have become serious weeds that sometimes dominate entire landscapes – and many of them are in fact far less palatable to livestock than the indigenous species were.

There are very few relatively undisturbed native grasslands left, but farmers have begun experimenting with restoration and replanting. The results are not always predictable, but it is already clear that indigenous species often produce better and more reliable grazing under a wider range of conditions and that the ludicrous sheep-stocking densities which helped to decimate so many species in the past have proven to be unsustainable on Australia's nutrient-poor, drought-prone, crumbly soils.

Not all revegetation is being done in the name of primary production, and an increasing number of projects are looking at ways of bringing back

Featherheads (*Ptilotus macrocephalus*)

a more complete range of grassland species. For the great majority of Australians who have never seen undisturbed grasslands the diversity of species is a revelation, because these include numerous plants with showy or even gaudy flowers, not just grasses. In spring, the massed heads of many types of daisies seem to burn hot patches through the green, while the plumes of Mulla-mullas and Featherheads (various *Ptilotus* species) add soft textures in pastel colours ranging from silver to warm pinks, and even purples.

A closer look reveals orchids and triggerplants, pink, white and lilac bindweeds, numerous types and colours of small-flowered native lilies, native geraniums with red or purple flowers, and creeping peas with flowers ranging from scarlet to blue and purple, or gold streaked with orange. On heavier soils which are waterlogged during wetter times of the year (often called 'grassy wetlands') even more striking flowers may dominate, from the surreal navy-blue starbursts of Blue-devils (*Eryngium ovinum*) to delicate mats of fairy-aprons (*Utricularia*), and a different range of daisies including the golden balls of drumsticks (*Pycnosorus* and *Craspedia*) held high above creeping carpets of beauty-heads (*Calocephalus*).

Rhodanthe chlorocephala, an everlasting daisy from Western Australia

Blue devil (*Eryngium ovinum*) prefers heavy, seasonally wet soils

We are unlikely to be able to restore the vast plains of native grasslands that early European settlers first lavished praise on then destroyed with ignorant enthusiasm, but many environmental groups are finding ways to rehabilitate smaller areas. The aim is to return or conserve as much biological variation as possible in relatively small spaces, where the species have a reasonable chance of success. The trial plots of grasses invariably include many non-grasses which have also suffered from overgrazing and the plough.

Craspedia variabilis, a distinctive daisy from seasonally flooded grasslands

If you're planning to recreate a patch of native grassland on

A nature-strip replanted with native grasses

your own property, there are many useful lessons from these ecologically oriented projects. First of all, use a selection of species propagated from your local area. These are of 'local provenance', and choosing local plants over random imports can be a significant way to help conserve the ever-decreasing gene pools of many indigenous plants. If there isn't an indigenous nursery nearby, it is often possible to apply for a permit to collect *small* amounts of propagating material from nearby remnants of grasslands. This won't be difficult to obtain if there are healthy patches of grasslands not too far away, but don't expect to be allowed to take anything from remnants that are so small they are endangered.

All grassland plants should ideally be propagated from seed to capture as much as possible of the genetic variability of local stocks. Conservation doesn't mean cramming the greatest number of different species into the smallest space possible, as it is better to have a larger number of a few plant species which can form small breeding populations. Thus, for a so-called nature-strip restored as a miniature grassland you may only be able to usefully include a half-dozen species, whereas for a small suburban block this number could usefully be doubled. For every grass species you use, you should include at least one other species of flowering plant though in smaller numbers. Your aim should be to recreate the complex living tapestry of natural grasslands, not just a sea of grasses.

Native lawns

Hemarthria uncinata: tolerant of drought, flood and salt ... and regular mowing

The basic design of most true grasses allows them to grow leaves from ground level, an adaptation to grazing by herbivorous mammals as the tip of the leaf may be eaten while the base will continue to shoot. Such grasses can be cut close to ground level to form a lawn: a dumbed-down grassland with a single dominant species. It is ironic that many Australians persevere with water- and fertiliser-hungry lawns of introduced grasses while in Europe and North America more natural 'meadow' plantings have long been fashionable. For these, mowing is delayed to allow a more diverse range of plants such as fritillaries, poppies and daisies the time they need to flower and set seed. Grasslands are the Australian equivalent, and offer at least as wide a range of colourful, flowering meadow/grassland plants as can be found in the entirety of Europe.

Most exotic lawn grasses are not ideal for Australian conditions as their short-cropped tops produce a correspondingly short and stunted root system, so they usually need watering to stay alive through even the mildest Australian summer. There are a few drought-tolerant introduced lawn grasses but, not surprisingly, these are by their very nature weeds which in many cases have already spread widely. Even these weedy species need fertiliser to keep them lush, and what escapes beyond the reach of their roots ultimately ends up being washed into waterways where it contributes to toxic algal blooms.

By contrast, native grasses are generally well-adapted to the nutrient-poor soils of our well-worn landscape, and selecting locally available species will pose no threat to the wild places and plants that still survive in your area. Wallaby grasses (*Rytidosperma*) are widespread across much of southern Australia, Windmill grass (*Chloris*), Blady grass (*Imperata*), Redgrass (*Bothriochloa*) and even low-growing *Poa* species have all been used as lawns, but there are lesser-known species with comparable or even better potential, particularly in difficult situations where few other plants will thrive.

Weeping grass (*Microlaena*) is pre-adapted to mowing, as it forms dense mats known as marsupial lawns in both shady and more open places when grazed by wallabies. In more tropical areas beard grasses (*Oplismenus*) can form lush carpets of springy foliage even on soils that may dry out for part of the year, while Mat grass (*Hemarthria*) and Salt couch (*Distichlis*) will tolerate saline soils as well as seasonal drought. The trick with these generally adaptable grasses is to not mow them too short, as that is an inappropriate aesthetic imported from much cooler and wetter climates where close-cropping the tops won't kill the stunted roots if no rain falls for a few weeks.

Many of these species don't like too much foot traffic or dog urine, and who shall blame them? It's an indication of their low nutrient needs, and just leaving the grass clippings to decay naturally into the ground should be all that's needed to keep them healthy in the long term. An ever-increasing range of native lawn grasses is available as inexpensive seed for larger plantings, or as plugs for situations where quick establishment is a priority. Regardless of their ultimate drought-hardiness, all these plants will probably need some watering over their first season, just as does any other type of lawn. At the other extreme, for wet places where mowing is usually impractical, there are many naturally low-growing sedges and related plants such as *Centrolepis*, *Eleocharis*, *Isolepis* and *Schoenus* that probably won't need any care or maintenance at all.

THE BUSH GARDEN ... AND BEYOND

Grasslands are open lands where trees and other taller species are generally sparse, needing more rain than the amount that falls in most years, or in some places killed off by too-frequent fires. But many grasses prefer life in more shaded places from bushland to rainforest and may be so abundant that they form a major component of the understorey. For example, Silvertop wallaby-grass (*Rytidosperma pallidum*) is often the most abundant plant in dry, open woodlands in south-eastern Australia, colonising even raw clay roadsides along forest fringes and yet it is often overlooked in even the most naturalistic bush garden plantings. And while various *Lomandra* (including the widespread and very variable *L. longifolia*) are regularly seen as specimens in open plantings, few gardeners seem to have registered that they may be the dominant species in diverse areas of natural bushland.

Though not everyone wants a bush garden, especially if you live near flammable natural bushland, these species are mostly amenable to more formal settings and make no objection to being planted among even the most exotic trees and shrubs from other parts of the world. Soft or hard, geometric or flowing, delicate or imposing: there is no difficulty in finding native grasses that will fit comfortably with the way you want your garden to look.

Rytidosperma pallidum

Lomandras grown as rock garden plants, Cranbourne Botanic Gardens, Victoria

And of course, there is also no reason why you can't replace all exotic grasses with indigenous species such as huge sawsedges (*Gahnia*) instead of invasive pampas grasses (*Cortaderia*), or make dense, soft-looking, low-maintenance native lawns in sun or shade. Towering speargrasses (*Austrostipa*) or tussock grasses (*Poa*) can be used to define avenues along garden paths and driveways just as well as any introduced species, and when they are past their best, the flowerstems can just be tweaked out to leave low rows of tidy, compact tussocks.

You can use native grasses to fill out the tiny front yard of your inner-city terrace with exotic-looking creations, to the bemusement of your neighbours, who will almost certainly be amazed to learn these species were once among the dominant plants of every suburb for many kilometres around. Not even a long, hot, dry spell in summer will affect the soft texture that many native grasses bring to a garden, and some including many swordsedges (*Lepidosperma* species) become even more colourful and vari-textured after a long spell of dry weather.

Lepidosperma concavum

Drought-tolerant plants

Even the most drought-tolerant species are likely to need some watering for their first year or two while roots establish, though this may only be once every few weeks during drier periods. The following groups include at least some drought-tolerant plants.

Anigozanthos, Arthropodium, Austrodanthonia, Austrostipa, Baloskion, Baumea, Bolboschoenus, Bulbine, Capillipedium, Carex, Caustis, Centrolepis, Chloris, Chorizandra, Conostylis, Cycnogeton, Cymbopogon, Cyperus, Desmocladus, Dichanthium, Dichelachne, Eragrostis, Eurychorda, Evandra, Ficinia, Gahnia, Gymnoschoenus, Hemarthria, Hypolaeana, Imperata, Johnsonia, Juncus, Kingia, Lepidosperma, Lepironia, Leptocarpus, Lomandra, Macropidia, Meeboldina, Mesomelaena, Neurachne, Notodanthonia, Pandanus, Patersonia, Poa, Spinifex, Themeda, Tremulina, Triodia, Vetiveria, Xanthorrhoea.

Lomandra patens in a bush garden setting

For these and many other reasons, native grasses have been enthusiastically adopted by some sectors of the gardening and landscaping communities, but it is disappointing that many of them formed their ideas of what grasses are and how they should be used long ago, with little change or innovation since. As a consequence, there is a tendency to overdo the use of a few well-known and easily mass-produced species to the point of tedium.

Common tussock (*Poa labillardierei*), Kangaroo grass (*Themeda*) and the various mat-rushes (*Lomandra*) are often stuffed into any situation where a grass-like plant is desired, regardless of whether they are likely to flourish, let alone look their best. In the case of mat-rushes this has led to a proliferation of new, named varieties of rather similar-looking and not necessarily exciting mounds of green, all of them (according to their tags) reputedly superb survivors in drought-prone situations. In reality, most mat-rushes are more likely to look a bit withered, sun-blasted or even sickly if they are left to struggle with drought and full sun on traffic islands, without even the occasional watering let alone any fertiliser.

There is nothing wrong with using such plants in moderation, or in conditions more closely approximating those they are adapted to. Some of them are very attractive when seen at their best. Common tussock can grow into a magnificent specimen in the permanently moist soils it prefers, such as near creeks and streams protected from the hottest sun, though it will also survive drought. Its parchment-coloured foliage can be striking even when it is semi-dormant – as long as it was *already well-established* before the first time it dries out. By contrast, the greener and finer-leaved species and forms of mat-rushes are woodland plants which prefer some protection from the hottest sun, and the bluer forms and species tell you by their waxy bloom that they are naturally adapted to hotter, drier and more open situations, needing the sun to bring out their best colour.

Perhaps the only widely planted native grass which has not suffered from overuse is Feathery speargrass (*Austrostipa elegantissima*), among the most beautiful and widespread of all native grasses. I have seen this planted in inland towns in the Victorian Mallee along with porcupine grasses (*Triodia* species), scrambling among banksias in Kings Park in Perth, and creating a soft and feathery surround for diverse indigenous plants in many other botanic gardens. Opening as slender and wispy, the

Native grasses used in a stylised planting in Geelong Botanic Gardens, Victoria

long heads expand as they mature into silken pennants which lock loosely together to form complex, silver mounds of foam.

There is no reason to stick to this handful of over-familiar grasses, however. Australia is more than just a country, it is a continent which stretches from the tropics to the Roaring Forties. The great variety of native grasses and related species potentially available to gardeners is increasing, and this book is only an introduction to their real diversity. Even the many and abundant rushes (*Juncus* species), familiar mostly to farmers as pasture weeds, look very different when they are grown in a

garden and are not constantly battered by cows, and in gardens the strongly upright habit of many species suits even the most formal plantings.

Among the true grasses, Foxtail mulga-grass (*Neurachne alopecuroides*) is a superb yet virtually unknown ornamental, with slender, vertical stems supporting masses of luminous blue-grey paintbrushes held in an open, upright fan. The low, rounded cushions of Buttongrass (*Gymnoschoenus*) are very different yet equally striking when in flower, with arching stems nodding high above the foliage, topped

A mixed planting of speargrasses (*Austrostipa* species)

with prominent buttons like oversized and slightly wilted fencing foils. Despite my earlier comments on the range of native plants that appear in nurseries in the guise of grasses, many of these such as the various cottonheads (*Conostylis*) are very grass-like indeed with their gunmetal-grey foliage silvered with fine fur – as long as you don't mind their seasonal displays of gaudy flowers. And who does?

Juncus sarophus in a coastal setting, northern Tasmania

Conostylis candicans

Flowering plants with grass-like foliage

None of the grasses, sedges or rushes have particularly colourful flowerheads. The plants included here are more distant relatives, some of them grown primarily for their flowers rather than their foliage.

Anigozanthos, Arthropodium, Astelia, Bulbine, Conostylis, Crinum, Dianella, Dietes, Diplarrena, Doryanthes, Dracophyllum, Helmholtzia, Libertia, Macropidia, Orthosanthus, Philydrum, Richea, Sowerbaea.

Dianella tasmanica in flower

WATERGARDEN PLANTS

Many grassy plants need seasonal or even permanent flooding if they are to look their best. The best known are the sedges, among the most ecologically important plants of wetlands and stream banks. Relatively few of the true grasses are as water-loving as sedges. The major exception is reeds, two widespread species of which dominate vast areas of wetlands worldwide, while in the north several species of ancestral rices (*Oryza*) cover comparably impressive areas on floodplains which dry out during the hot season.

Aquatic sedges range greatly in size and habits. They include some fine specimens for any size of water garden, from the tidy, mounded tussocks of Tassel sedge (*Carex fascicularis*) to the gigantic and sprawling Leafy twigrush (*Cladium procerum*). Some will trail in moving waters, swirling elegantly with the current, while others form lawns or upright carpets in the shallows, growing so densely they smother competing

Baumea preissii in Kings Park, Western Australia

The upright habit of *Cycnogeton procerum* creates an attractive contrast with soft-textured Lake water-milfoil (*Myriophyllum salsugineum*)

weeds. Fully adapted to Australian conditions, most aquatic sedges will also tolerate some degree of drought. The twigrushes (*Baumea*) especially will thrive in water throughout the year, yet still look healthy and green after months exposed on apparently dry soils.

Other grass-like families which are primarily aquatic, whether they prefer deeper waters or the shallows, include the familiar cumbungis (*Typha*), which many people still call bulrushes. Although selected forms can be highly ornamental and not difficult to manage in a pot in the smaller water garden, cumbungis are invasive plants which should not be deliberately introduced into dams: there are many other grassy plants which provide better habitat and are more easily managed. Water ribbons (*Cycnogeton*) may grow in huge numbers in shallow waters which dry out at times, and the intriguing Woolly frogmouth (*Philydrum*) and its close but even more striking relative the Stream lily (*Helmholtzia*) are always found in or near water. Tassel cordrush (*Baloskion tetraphyllum*) in its many forms is the most widely planted grassy wetland plant of all, and some forms can form a huge tussock with soft and feathery plumes that can reach well over 3 m high. Many other native cordrushes are also water-lovers.

Plants for watergardens and wet places

Most genera here include at least some species that grow in permanently damp to wet places. Check the individual species entries in the encyclopedia for more detailed information on their preferred growing conditions and tolerances.

Lepidosperma elatius (at left) and *Juncus procerus* form a pleasing textural contrast in this farm dam repurposed as habitat

Amphibromus, Baloskion, Bambusa, Baumea, Bolboschoenus, Carex, Chaetanthus, Chorizandra, Cladium, Cycnogeton, Cyperus, Distichlis, Echinochloa, Eleocharis, Fimbristylis, Fuirena, Gahnia, Glyceria, Hanguana, Hemarthria, Hypolaeana, Isachne, Ischaemum, Isolepis, Juncus, Lachnagrostis, Lepidosperma, Lepironia, Leptocarpus, Leptochloa, Oryza, Patersonia, Philydrum, Phragmites, Pseudoraphis, Rhynchospora, Schoenoplectus, Schoenus, Scirpus, Sparganium, Tremulina, Triglochin, Typha, Vallisneria, Xyris.

POTS AND SPECIMENS

Most true grasses don't make good pot specimens as they need reasonably dry conditions for at least part of the year yet will die if allowed to dry out completely in a pot. By contrast, the deep roots of such species will always be able to find enough moisture to get by when they are established in the ground. The balance can be difficult to achieve when planting in a pot, though I have had good (if short-lived) results with Feathery speargrass (*Austrostipa elegantissima*), with its airy, open flowerheads, though this handsome plant is fairly ordinary-looking when it has been tidied back to a green tussock. On the other hand, some groups such as rushes seem almost custom-designed for pots with their various neat and predictable non-running habits.

The best of the true grasses for pot culture come from rainforest areas, as these species are adapted to constant soil moisture and overwatering is unlikely to become a problem. They include the beard-grasses (*Oplismenus*), which are also known as basket grasses for their cascading habit when grown in a hanging basket. Although these shade-tolerant species are occasionally grown in the warmer parts of Australia, they will thrive as a potted plant much further south if protected from frost. Native bamboos and several of the bamboo-like rainforest grasses will also thrive in a well-watered pot, with the exception of the climbing vine *Muellerochloa* which is temperamentally unsuited to this purpose, but that species is not readily available in any case.

Among the sedges, there are many water-loving species which aren't worried by months of drought when grown in soil,

Lepironia articulata and *Curculigo capitulata* grown as potted specimens

These wild-collected grasstrees appear healthy but are less likely to thrive than seed-grown plants, particularly the two that are already in flower

and these will usually adapt well to pot culture. They include such strikingly upright and formal-looking plants as Spear-sedge (*Lepironia*). Sedges originating from drier places have deeper root systems and rarely thrive in a pot. Other groups which look attractive as potted plants and seem to thrive under these conditions include many of the rushes and cordrushes, mat-rushes and even the flax-lilies, though these may not flower as freely as they would if planted directly into the garden.

Grasstrees are still often sold as 'established' specimens in fancy pots, but these should be avoided. They are unlikely to live more than a few years in a container as this is a very different environment from their natural bushland habitats. Many such specimens were ripped from the ground, so their best chance for survival is to establish an extensive new root system into deeper soils before the starchy reserves in their trunks run out. If you are hoping to grow grasstrees in a pot, start with seedlings of any of the new hybrids, which are faster-growing than wildlings. However, they will need regular watering during dry weather, creating ideal conditions for a range of diseases that grasstrees have little resistance to. The deadliest of these is cinnamon fungus.

Plants suited to containers and pots

These groups will all remain healthy in pots and some may even flower more vigorously when root-bound. Only species which make particularly attractive specimens have been included here. There are quite a few other species that would also be worth experimenting with if they particularly appeal to you.

Ficinia nodosa

Arthropodium, Baloskion, Bambusa, Chloris, Cordyline, Crinum, Curculigo, Dianella, Dietes, Diplarrena, Evandra, Ficinia, Fimbristylis, Gymnoschoenus, Helmholtzia, Isachne, Isolepis, Juncus, Lepironia, Libertia, Oplismenus, Schizostachyum, Sowerbaea, Xyris.

Using local grasses in the garden – if you can

It isn't hard to understand why it is worth conserving as much local variation as possible in local populations of most grassland plants, yet some nurseries still promote the idea that provenance isn't important for grasses because they are pollinated by the wind. Yet even wind-blown pollen doesn't drift all that far before it settles or dries out. There are also considerable genetic differences between local populations of common grasses which may even be obvious to the naked eye. Consider *Themeda triandra*, perhaps the most widespread native grass, found not only throughout Australia but also abundant through warmer parts of Asia and much of Africa. Within Australia we call this plant Kangaroo grass, though unsurprisingly the name doesn't seem to have caught on in the many other countries where it is found.

Common tussock (*Poa labillardierei*) softens the edges of a gravel driveway

The genetic makeup of Kangaroo grass must vary considerably over this great range no matter how hard the wind tries to spread its pollen around, and even within Australia there are many distinctive populations. These range from the tall, upright forms of south-central Queensland (which are sometimes over 2 m in height) to sprawling, green-grey miniatures of unknown origin such as the cultivar 'Mingo', as well as many others that can be recognised at a glance. There must also be less

Diverse regional grass-like plants planted in raised beds, Geelong Botanic Gardens, Victoria

obvious differences hidden in the genes of localised populations of many different grasses, which could easily be submerged by widespread use of seed imported from elsewhere.

For all such widespread grasses, it makes good ecological sense to stick to your local forms. These are often surprisingly easy to obtain from any garden centre which takes the trouble to offer a good selection of local plants. 'Staying local' should also be your guiding principle for groups of look-alike species including many rushes (*Juncus*), wallaby grasses (*Rytidosperma*), speargrasses (*Austrostipa*) etc., as even botanists may have trouble identifying the differences between a range of closely related species from far-flung parts of Australia. For gardeners there should be no problem – if your local species look much the same as the photo you have admired of a plant that grows 1,000 km away, why not grow just the local species or form and support the nurseries that put work into preserving your local bushland?

Even so, gardeners don't have to be fanatical and plant just those few species and forms of grasses and related plants which grow (or used to grow) within a few kilometres of where they live. Many of the native grasses and related plants described in this book are regional specialties which won't make any difference to local bushland plants, and there is little or (more probably) zero chance that they could establish in nearby bushland, let alone become weedy outside their natural range.

Examples of grass-like plants which can be used widely without causing problems include Palm grass (*Curculigo*), which is only likely to thrive in well-ordered, sheltered, frost-free gardens, even though it may be planted 1,000 km south of its natural range. Nearly all cordrushes also have very specific needs in the wild. We don't even know how to encourage them to self-seed, and as these are mostly either male or female you could always grow just one sex if you are worried that escapees could become a problem. At the other extreme, native lemongrasses seem to tolerate anything from long-term drought to months of torrential rain so these are already widespread in both eastern and western Australia. So are most bluegrasses, various clubrushes, sawsedges and swordsedges, mat-rushes and spikerushes, and many other striking plants described in the encyclopedia which makes up much of this book.

Two plants of Velvet tussock (*Poa morrisii*) are enough to transform this meandering path beyond the ordinary

3

Growing, propagating and maintaining native grasses

The foliage of kangaroo paws such as *Anigozanthos* 'Big Red' is attractive throughout the year, but the plants are stunning when in flower

THERE ARE FEW MYSTERIES WHEN it comes to growing the grasses and related plants described in the next section of this book, as they are mostly adaptable species that tolerate a wide range of soils (or water conditions, in the case of the aquatics), and even drought and heat once they are well-established. They are also virtually pest-free if grown under conditions close to those they prefer in the wild. Any apparent pests that do appear are most likely to be indigenous creatures that cause little damage, including some very attractive and uncommon moths that won't be attracted to your garden by any other plants.

PROPAGATION

Although the emphasis in this section is on some of the more difficult species, most grass-like plants are easily raised from seed, and although the seedlings won't always come out identical this is the way some of the more interesting new variations arise. However, the seed of many wild grasses needs a dry dormant period, usually of several months, before it will germinate and for some sedges you may need to wait a year or more. Seed won't necessarily come to harm if it is planted immediately after harvesting, but the longer it remains in well-watered soil the greater the

Seedlings of *Poa labillardierei* from the same parent plant give an idea of natural variation

likelihood that weeds will invade the trays or pots, and small seeds may be removed by ants before they sprout. In cooler areas, the best time to plant out seed is spring; in the tropics you should plant not long before the wet season sets in, to take advantage of natural growing patterns.

Many grass-like plants are easy to produce from divisions, or from sections of the runners in some cases, as these will usually put out new roots even without the use of rooting hormones. This is fortunate in the case of twigrushes and other groups that no one has succeeded in germinating in any quantity, even though their seed may be formed in good quantities and looks as if it contains viable embryos. More distant relatives of the grasses, including kangaroo paws, mat-rushes and the smaller lilies, are mostly easy to propagate by division but time this action for when the plants have finished flowering and are actively growing new roots and foliage instead of putting their energy into developing seed.

Unfortunately, some of the most striking sedges can't be reliably raised from seed, and also resent transplanting or division. If enough of their deep root system is dug up when transplanting, they may gradually re-establish a complete root system in a pot if left undisturbed for a year or more. In turn, established potted plants are more likely to survive splitting into several separate pieces later, because it is easier to lift out

Gahnia radula, a handsome sedge that rarely sets seed and is difficult to propagate by division

the entire root system and tease it apart, rather than cut through the roots indiscriminately. Once again, you will need to leave the freshly separated divisions in their new pots for some time while they recover and are fit to be planted out or multiplied further.

Many cordrushes are also difficult or even impossible to raise from seed, some because they rarely set any, and what little is produced is often short-lived. In the wild the seed of these species would germinate rapidly, and would thrive only in ideal conditions such as after fire. Even some of the wetland cordrushes seem to need fire to clear competing vegetation and to stimulate seed germination. A further complication is that nearly all cordrush plants are male or female, so it is essential to have both sexes growing close together or there won't be any seed formed at all. As most nursery-propagated plants are grown by division, it may not always be easy to locate plants of both sexes for seed production.

Most cordrushes don't have well-developed rhizomes, so propagation by division will only work reliably for a few of them. Timing the division for when the plants are starting to produce new roots and rhizomes (usually in spring) is critical. Plants that are already well-established in pots are the easiest to divide, as they can regularly be removed from the pot to check for new root and shoot growth. As soon as new shoots appear under the soil, the plant is growing actively and is more likely to recover from the injury of cutting into separate plants.

SOILS, FERTILISERS AND SALTS

Like many other native plants, grasses mostly come from nutrient-poor soils and some don't enjoy high levels of nitrogen and phosphorus, though it is surprising how many will grow more vigorously with a handful of

Ficinia nodosa thriving in nutrient-poor sand just above the high-tide line

blood-and-bone to start them off. Such plants may grow slowly in the wild, but it doesn't mean they won't respond vigorously to a suitable range of nutrients. If in doubt about fertiliser requirements, you can use the so-called 'native' fertilisers formulated for sensitive species such as banksias, though these are synthetic blend of salts and may cause unnaturally soft and fungus-prone foliage. In any case, most native grasses are fast-growing, so a single feeding when they are first planted will usually give them all they need to grow into a substantial plant, after which they should be left to mature at a more natural pace.

Many inland species will tolerate a fair degree of salinity, and even Kangaroo grass can be grown in coastal areas where there is enough seasonal rainfall to occasionally flush the gradual buildup of salt in the soil, caused by sea spray. Plants which grow naturally along inland waterways are invariably salt-tolerant to some degree, as water flowing through these low-lying landscapes is rarely flushed by heavy rains. And of course coastal species such as spinifex grasses are undisturbed by salt spray, or even the occasional king tide.

Plants for coastal gardens or saline soils

These genera include species that are found naturally in slightly to very saline conditions, from the upper reaches of estuaries to coastal dunes with frequent drifts of sea spray.

Baumea, Cycnogeton, Dichelachne, Dietes, Distichlis, Ficinia, Fimbristylis, Gahnia, Hemarthria, Ischaemum, Juncus, Lachnagrostis, Lepironia, Leptocarpus, Phragmites, Poa, Schoenoplectus, Spinifex, Themeda.

Juncus kraussii mostly grows in tidal areas

WATER AND DROUGHT, SHADE AND LIGHT, FROST

No matter how drought-tolerant they may be, all newly planted grasses must be watered adequately for the first year or so until they have developed extensive root systems. Watering doesn't need to be done

Common tussock and various speargrasses – attractive both growing and flowering, but also dried out after months without rain

often, and you can rely on the long and slender leaves of grasses to roll up into elongated cylinders as a warning if they are feeling moisture stress, long before they are in any danger of drying out seriously. It is better to give a recently planted grass a slow, deep soaking with 1–2 litres of water every couple of months over its first hot season, than to hose it lightly so the soil is only wet near the surface and the moisture evaporates before the plant has a chance to use it.

Although you shouldn't need to water drought-tolerant plants once their root systems are established, where you plant them can affect the way they grow and look in the long term. In full sun, grasses tend to grow compactly and you may need to remove dead foliage more often than when they're planted in lightly shaded situations – though there is a certain

Shade-tolerant plants

The groups included here all have species which are often naturally found in shaded places, though many others in the encyclopedia will also grow well with less sunlight.

Baloskion, Baumea, Carex, Crinum, Curculigo, Cyperus, Empodisma, Fimbristylis, Flagellaria, Gahnia, Helmholtzia, Isachne, Isolepis, Juncus, Lepidosperma, Lomandra, Microlaena, Oplismenus, Philydrum, Pseudoraphis, Schizostachyum, Tetrarrhena.

Isolepis cernua is upright grown in sun and becomes more open and weeping in shade

Frost-tolerant plants

Many grasses will survive an occasional frost to -2–3°C without serious damage, some of them will tolerate much lower temperatures, and many will thrive even in regularly frosted areas as long as they are planted in the shelter of other species or against warmer, north-facing walls and embankments. As usual, check the encyclopedia for more detail on the individual preferences of particular plants.

Dianella tasmanica 'Cherry Red'

Anigozanthos, Arthropodium, Austrostipa, Baloskion, Bulbine, Carex, Chloris, Chorizandra, Conostylis, Cycnogeton, Cyperus, Dianella, Dichanthium, Dichelachne, Diplarrena, Eleocharis, Eragrostis, Gahnia, Gymnoschoenus, Imperata, Joycea, Juncus, Lepidosperma, Lomandra, Microlaena, Neurachne, Patersonia, Phragmites, Poa, Richea, Schoenoplectus, Themeda triandra, Triglochin, Triodia, Typha, Xanthorrhoea.

austere beauty to many dry grasses. With increasingly deep shade (especially with protection from the hot, afternoon sun) and higher moisture levels overall, the foliage will usually remain greener and may spread more widely. Consider the upright stems of Knobby clubrush (*Ficinia*) grown in an open area, in contrast to its distinctly weeping habit in more sheltered situations.

TRIMMING AND TIDYING

Grasses and all their relatives grow very differently from trees and herbs, with each leaf growing from the bottom outwards or upwards. This means that if you try to tidy a grass by cropping the tips to achieve a particular shape, you are likely to create a peculiar eyesore as some leaves continue to grow longer, while older ones remain the same. Regardless of age, each leaf will finish in a ragged, torn and often shrivelled tip – and these frayed remains will stick out at random distances from the centre of the plant, giving a battered and unwholesome effect.

Although you can't prune grasses, it isn't hard to select and plant each species so that its final habit and shape accords with the effect you want. This kind of information is included on the descriptive tag that comes with the pot for many named cultivars. You can still clean the plant up when required, in two very different ways.

The tidiest and least controversial method is to tug out dead leaves you find objectionable, as these may completely shroud the healthy, living part if allowed to build up too much. Dead leaves and stems usually come out fairly easily, even by the handful in many cases, though your tactics may need to be adjusted for different species. Some leaves come off with the slightest tug, while for tougher leaves and stems you will need to reach in closer to the centre of the plant to get a firmer grip – preferably while wearing a glove as even smooth-looking leaves or stems may cause unpleasant, long-lasting cuts when (not if) they slip through your hand.

The other alternative is firestick gardening, where tussocks with an excessive accumulation of dead leaves are simply set on fire at an

Lomandra longifolia after 'pruning' – not a good look!

Poa tussocks after a poorly controlled burn – another unattractive 'tidying' measure

appropriate time of year – definitely not in summer. Although an effective way of disposing of long-neglected dead leaves, this shortcut adds to the carbon dioxide levels in an already overloaded atmosphere, though admittedly very little compared to the amount you add by driving to your nearest garden centre. The greatest disadvantage of fire as a pruning tool is that it may kill the plants if it becomes too hot, so keep a hose or a pressure sprayer handy and deliver a fine mist occasionally if temperatures become excessive.

How long do native grasses live?

Grasses and their relatives are a wildly variable group, ranging from annuals which may only live for a few months, to grasstrees which live for many centuries. As these plants don't have a woody trunk there is no easy way to guess their age, such as by counting growth rings. We do know that ancient rings of Porcupine grass (*Triodia scariosa*) must have taken centuries to reach the stage where their centres have not only

An ancient *Triodia scariosa*. The centre of this venerable specimen probably died centuries ago

died, but have decayed away completely even in the arid inland climates where such plants are mostly found.

Yet even grasses we would intuitively regard as short-lived may perhaps live longer than the humans who plant them. During 2001, in the course of works at Geelong Botanic Gardens, a section of lawn being converted to new beds was fenced off to protect visitors from earthmoving machinery. The fences also prevented mowing, and within months the rough, unwatered lawns sprang into life and turned out to be Kangaroo grass, speargrasses and wallaby grasses, with a few other native grassland plants – and almost no introduced grasses among them!

Old-timers from around the area recalled that the lawns hadn't always been mown. For close to a century they had been regularly cropped for hay, usually before the seedheads had a chance to ripen. The golden-brown heads they recalled were mostly of Kangaroo grass, which had not managed to produce seed in all that time, so the plants flowering in 2001 may have lived the best part of a century despite being constantly cut down. Kangaroo grass is just one of several native grasses whose grain potential is being considered. It could be grown as a perennial crop that doesn't need replanting – and which could help bind our disappearing topsoils instead of destroying them through ploughing as conventional grain-growing does.

DROUGHT AND FIRE

Many of the disturbing predictions about the effects of global warming seem to be coming true, with increased flooding in the north of Australia, hotter and drier summers in much of the south, and water restrictions that will increasingly limit what we can grow. Garden writers usually regard this as a reason to grow Mediterranean-climate plants, even though many introduced plants that thrive in the warmer summers we are experiencing are potential weeds.

Many indigenous grasses are very tolerant of drought, however, and may be even more attractive when their leaves and seedheads have turned to bronze than when they were fresh and green. Such plants may look dead but this is just a part of their normal lifecycle, and there is no reason to cut or trim them unless they pose a fire risk or are close to the house. Save any tidying up until after the onset of the next rains, when the old foliage will usually collapse. The plants will spring into new life over the next few months – or even weeks, in some cases.

Drought comes with fire, and the worst fires invariably happen after months with little or no rainfall. Many native grasses grow in fire-prone places, and at least some are reliant on fire to control competing vegetation such as trees and shrubs that would eventually shade the

Grasstrees at peak flowering, many months after a fire started by lightning

grasses out. Buttongrass (*Gymnoschoenus*) is a good example, flowering vigorously for a few years after a fire has passed through. The flowering of grasstrees (*Xanthorrhoea*) as a response to fire is more familiar, because the exposed, ash-rich soils left by bushfires are an ideal seed bed, giving the seed a chance to germinate and establish before other plants have the chance to move in.

The slender leaves of all grasstrees catch alight easily, setting up an updraft that helps to cool the crowns as the fire passes through, so these will usually retain a protective thatch of leaves except after the hottest blazes. Within weeks the grasstrees will be growing again, with the stored sugars in their trunks fuelling rapid growth of the tall, poker-like flowerheads, which may stretch several centimetres with each passing day. The following spring the pokers turn white with millions of tiny flowers, like lilies without petals, providing a nectar-rich, fragrant feast that draws huge numbers of pollinators from insects to honeyeaters.

This degree of adaptation has direct implications for gardeners in bushfire-prone areas of southern Australia, which are distinguished by some of the worst wildfires in the world – including where I live near the Great Ocean Road in south-western Victoria. Although most grasses and their relatives don't carry a high fuel load compared to native forests, they

Burning off *Gahnia sieberiana* around our swimming dam

will flare rapidly on a blowout day, generating a lot of heat for a short time. This is not necessarily a problem if the plants are at a safe distance as the heat reduces rapidly as soon as the fuel is mostly gone, but you don't want dry grasses too close to your house in summer.

Fire-adapted grasses don't have to be dry to be dangerous. Some will flare readily even while apparently lush and green, and will regrow over the next two or three years. The most volatile types include grasstrees, some swordsedges (*Lepidosperma*) and particularly the sawsedges (*Gahnia*). The heat from such relatively small fires is localised as the amount of fuel available isn't much compared to burning bushland, and the most intense heat is soon over, but such plants aren't desirable close to the house. The photo shows a small burn-off at our swimming dam, at the end of a damp spring day in 2012. Although most of the surrounding vegetation was undamaged we have since removed most of it, and though we have left many vigorous sawsedges (see the opening image of Chapter 2) these can be burnt off with much less drama whenever a serious fire season looks to be developing.

Edible and useful species

Swordgrass brown butterfly (*Tisiphone abeona*) – strictly speaking, this should be a sawsedge brown!

Many native grasses, sedges and rushes are significant sources of food for diverse native animals ranging from insects to birds; for example, many butterflies need specific grasses or sedges for their caterpillars to feed on. As the true grasses include all the major food crops (wheat, rice, barley, rye, sorghum and millets, etc.) that make up more than half of the food we humans eat, it is not surprising that the grain potentials of many other grasses – including a fair few Australian species – are attracting interest in the increasingly unpredictable and adverse conditions of a world heading rapidly towards climate crisis.

Oryza rufipogon grows naturally through much of northern Australia

In terms of potential contributions to the development of significant grains, Australia is a long way from being a backwater. For example, the greatest genetic reservoir of the primary wild ancestor of rice (*Oryza rufipogon*) is in the seasonal wetlands and floodplains of northern Australia. Through most of its original range in South-East Asia, the wild ancestors of the cultivated grain have long been overwhelmed by rice terraces that are beautiful to look at but are not very genetically varied. Sorghum is an unfamiliar crop to most Australians even though it is the fifth most significant grain crop worldwide, yet our native species are estimated to include 90% of worldwide genetic variability for this genus. This offers potential for breeding perennial varieties that won't need ploughing and replanting every year, thus reducing soil erosion.

Developing seedheads of *Triodia irritans*

Future potential crops among native grasses include Weeping grass (*Microlaena*), seen by some as a potential gourmet crop when the grain is parched after the fashion of North American Wildrice (*Zizania*). However,

Edible tubers of 'Chinese' water chestnuts, a widespread native in northern Australia

you shouldn't expect to both use this as a lawn and harvest a substantial crop, as the more often it is mown the less seed it will set. Kangaroo grass (*Themeda*) is another possible food crop that, as it is a surprisingly long-lived plant, could also form the basis of a perennial grain cropping system. Even the aggressively self-protecting, spiny-tipped spinifex grasses (*Triodia*) of arid areas have been harvested for seed.

Beyond true grasses other species with high-quality edible parts are described in the encyclopedia.They include a few particularly productive sedges such as *Bolboschoenus* with substantial tubers up to the size of a walnut, native forms of 'Chinese' water chestnut (*Eleocharis dulcis*) which cover huge areas of wetlands in northern Australia, and the diverse water ribbons (*Cycnogeton*) with their sweetish, starchy tubers sometimes produced within a year of planting. Even the weedy cumbungis (*Typha*) are prolific sources of bush tucker, offering nutrient-rich pollen, fine-grained root starch and crisp new shoots which have been compared to a rather bland form of asparagus.

4

Encyclopedia of Australian grasses, sedges and rushes

WELL-STOCKED NURSERIES AND GARDEN CENTRES often have a remarkable range of plants lumped together as native 'grasses', quite a few of them related so distantly (if at all) that the best you can say about them is that their leaves are relatively slender, and they might *almost* pass as grasses when not in flower. But a line must be drawn if this book is not to double in size, fluffed up with plants such as daisies. My approach has been to stick to species that are 'reasonably' closely related to the true grasses, sedges, cordrushes and others that make up the bulk of the entries, or that are widely assumed to be grass relatives and have common names to match. This subterfuge has allowed me to slip in giant heaths (*Richea* and *Dracophyllum*) as they are often called grasstrees in their native ranges.

Many plants related to lilies, including a few from the asparagus family, are also described, but with the stipulation that if their foliage is not anything special and they are mostly grown for spectacular flowers, or they are tricky to maintain in the garden (for example, *Blandfordia* fits both these categories), then they are not really grass-like in any useful sense. However, I have included the kangaroo paws (*Anigozanthos*) because many of these are grass-like for most of the year, and they are usually included with grasses in most nurseries.

When using the encyclopedia, keep in mind that the photos are just a selection of the species you are most likely to encounter in the larger cities where the great majority of Australians live. Often there are very similar-looking local species appropriate to your area, which will grow and look much the same as plants from 1,000 km away. Some local variants aren't hard to find (indigenous nurseries are the best source for these), and these should always be used in preference to similar-looking imports from elsewhere which may have weed potential.

Amphibromus (Poaceae)

Swamp wallaby-grasses form extensive, elegantly weeping stands in seasonally flooded places, and are a significant habitat for many species of small, winter-breeding frogs better known as froglets in southern Australia. These grasses look superficially similar to Austral sweetgrass (*Glyceria*) but the spikelets of swamp wallaby-grasses have a long awn, so each spike looks a bit like a miniature head of wheat (see the comparison of these two in the earlier introduction to plant families). The most widespread and readily available species is Common swamp-wallaby grass (*A. nervosus*), found from Western Australia to New South Wales

Amphibromus nervosus

and Tasmania, with pale-green flowerheads. Dark swamp wallaby-grass (*A. recurvatus*) is also sometimes grown for its purplish flowerheads. Most of these plants reach around 1 m in height (less, in drier conditions), sprawl outwards to a similar extent as the seedheads mature, and are readily grown from seed sown onto moist to wet soil.

Anigozanthos (Haemodoraceae)

Nurseries usually group kangaroo paws with grasses on the basis of their foliage, relatively slender in some of the species though broader and more iris-like in others. However, if you plant kangaroo paws because you think they look appropriate among other grassy species, you will be in for a surprise when they flower. These are among the gaudiest and heaviest-flowering of native wildflowers and are the basis of a substantial industry with an ever-changing range of garden selections and hybrids. The 'Bush Gems' series includes many smaller plants that work well in smaller gardens, but some taller selections such as 'Yellow Gem' look distinctly top-heavy when in flower.

Many of the more recently developed forms are difficult to grow except in areas with reasonably warm and dry seasons, and rapidly succumb to fungal diseases or frost in cooler, damper climates. They

Anigozanthos flavidus green form

cover the gamut of fiery colours from vivid yellows and oranges to reds and crimsons, and range in size from dwarf selections only 50 cm high to giants which tower over many gardeners. The wild forms are generally more subdued in colour than hybrids and look more appropriate in a garden of grasses because they won't overwhelm other plantings when in flower. Tall kangaroo paw (*A. flavidus*) is particularly tolerant of cooler, damper areas. Green and red forms of this species thrive along the more open sections of our driveway, tolerating wet winters, sometimes very dry summers, and occasional heavy frosts.

There is little point in attempting to describe all the named cultivars of kangaroo paws. New cultivars become available almost every year and not all of these will be available five years later, replaced by newer hybrids in this high-turnover industry. Both species and hybrids need good drainage, though in drier climates they will look greener and flower better if given an occasional soaking. Propagation of named varieties is by division in late spring, but some of the species can also be raised from seeds.

Arthropodium (Asparagaceae)

Vanilla lily (*A. milleflorum*) is a tuberous herb forming grass-like tussocks which may be upright in sunlit conditions, or somewhat weeping in more

Arthropodium milleflorum with flowers inset

shaded conditions. Found along the east coast from southern Queensland and as far west as South Australia, and as high up in altitude as the ACT, it is easy to grow in most soils but may die down to dormant tubers in dry conditions. It looks most natural massed in a woodland setting but is also attractive as a potted specimen. It is easily grown from seed planted in autumn and left out in cold, wet conditions, or clumps may be divided.

The masses of faintly scented lilac flowers appear from late spring into summer, looking superficially like dainty fuschia blooms with furry filaments projecting below, and are held above the foliage on stems to 1 m high. These attract diverse insects, and as a bonus the tubers can be eaten raw or roasted if you decide to thin them out. See also *Dichopogon*.

Astelia (Asteliaceae)

Astelias are an interesting group of lily relatives with dark-green leaves, sharply pleated along their length and densely furred with silver-white below. Tall astelia (*A. australiana*) from southern Victoria grows in deep mountain gullies so wet that the surrounding plants are mostly ferns, often so shaded they rarely flower even if their 2 m leaves manage to rise above the dense undergrowth. Although it has been grown in more open gardens (where it is much more likely to set seed), this uncommon species

Astelia australiana with details of the upper leaf

is now protected and has only rarely been available from specialist growers.

Fortunately, the foliage of Pineapple grass or Silver astelia (*A. alpina*) is just as strikingly contrasting above and below, forming sprawling tussocks to ~30 cm high with small heads of greenish flowers, followed by red berries. Widespread on boggy ground in alpine areas of south-eastern Australia and Tasmania, this species is abundant through much of its natural range. Though they look similar, mainland plants are regarded as a different variety, *A. alpina* var. *novaehollandiae*. Male and female plants growing together are needed to set seed, which doesn't seem to germinate freely, but larger clumps aren't difficult to propagate by division. Tolerant of freezing and short periods of heat, Pineapple grass prefers moist soils though it will apparently survive occasional low-intensity fires.

Astrebla (Poaceae)

Mitchell grasses (*A. pectinata* and several others) are long-lived, tussock-forming grasses with slender seedheads rather like a scrawny head of awnless wheat, and are still widespread across many parts of inland Australia. As far as appearance goes they aren't likely to take the gardening world by storm, but they have considerable potential as drought-tolerant perennial grains. Curly mitchell grass (*A. lappacea*) at least was harvested and stored, sometimes on a substantial scale, by Aboriginal peoples. These grasses are noted for their tolerance of alkaline soils and moderate grazing, as long as there is enough summer rain to allow them to recover.

Austrodanthonia – see *Rytidosperma*

Austrostipa (Poaceae)

Speargrasses were previously lumped with *Stipa* but form a distinctive Australian group of their own. They are among the most widely used and decorative native grasses in gardens, growing on a wide range of soils and needing little fertiliser. They are also often the dominant grasses (sometimes in mixed groups of two or three species) in dry bushland, remaining common along roadsides where most other native vegetation has long disappeared under a sea of introduced grasses.

Most speargrasses are more-or-less similar-looking tussocks forming a mound 30–60 cm wide with time, but their flowerheads are varied. Their common name comes from long, spiked seeds which readily impale socks and even skin, so be warned and don't plant them too close to places you like to walk. On the plus side, the seed is easily germinated after a dry resting period (many months in some cases), or if you want large plants sooner the tussocks can be divided in spring.

The most widely familiar species is Feathery speargrass (*A. elegantissima*), found across most of southern Australia. It is planted as a soft and feathery ornamental from Kings Park in Perth to the gardens of Melbourne and Sydney, and can be seen in inland areas where it may be among the most conspicuous native grasses in mallee scrub. The plant in flower forms a pendulous yet feather-light mass and a wild stand is a spectacular sight in a strong wind. It will scramble through bushes and form mounds up to 2 m high, or spread out just as widely.

The bamboo-grasses of New South Wales and subtropical Queensland are well-named with their fine, stiff stems forming open clumps, topped by ethereal, fine-seeded flowerheads. Slender bamboo-grass (*A. verticillata*) holds masses of delicate pink-green seeds up to 1.5 m high on stems that shift and bow with every passing breeze. Stout bamboo-grass (*A. ramosissima*, better known overseas as 'Pillar of Smoke') is essentially a taller, thicker-stemmed version that can reach over 2 m. Both species are drought- and frost-tolerant but are relatively short-lived. Older specimens are easily pulled up after several years as their root systems are shallow. They will usually be replaced by seedlings from the disturbed soil, which suggests they have weed potential outside their natural range.

Austrostipa elegantissima among *Poa labillardierei*

Austrostipa gibbosa

Austrostipa mollis

Austrostipa scabra flowing with the wind

Most other speargrasses don't reach much more than 1 m in height even in flower and are fairly upright, holding their young flowerheads close together like miniature clumps of pampas grass. The heads become more open and lacier in appearance as they mature. These are the most frequently available species, yet even some of the finest and most widespread including *A. mollis* and *A. semibarbata* don't have widely accepted common names and are simply referred to as 'speargrasses'. *A. scabra* is a smaller and more refined version of these taller species and looks particularly fine when tossed by a strong wind.

Austrostipa verticillata

Baloskion (Restionaceae)

This is perhaps the most common and ecologically important group of cordrushes in eastern Australia. It was formerly known as *Restio*, which is now regarded as an exclusively southern African genus. Tassel cordrush (*B. tetraphyllum*) is the most widely cultivated species and is found from Queensland to Tasmania. It is among the best known and most variable of all Australian grass-like plants, with numerous wild and garden forms available. *B. tetraphyllum* subspecies *tetraphyllum* is the tall form found in Victoria, South Australia and Tasmania, where it often grows in large groups in seasonally flooded areas, including along riverbanks. This southern form is usually a large and fairly upright tussock to 3 m or more, spreading almost as widely at the top while the bare lower stems remain tightly packed. The brown bracts at the nodes along the stems give a

Baloskion tetraphyllum subsp. *meiostachyum* near Jervis Bay, New South Wales

Baloskion australe

Baloskion tetraphyllum subsp. *tetraphyllum* in the Grampians (Gariwerd) National Park, Victoria

somewhat bamboo-like effect, while the fine foliage looks feathery from a distance.

From around the Victorian–New South Wales border northwards, the eastern coast subspecies *meiostachyum* is the dominant form. It is usually more compact and bushy-looking and its lower stems are less exposed, so the overall effect can be even more open and feathery than the southern forms. This subspecies also seems more variable. Smaller forms barely reach 1 m, so they look like some kind of shrub rather than a cordrush, but plants twice this height and width are not uncommon and some of the more attractive garden forms have been selected from these. For smaller gardens the 'Green Wedge' form is particularly fine, a densely growing plant no more than 60 cm high (smaller in pots) and sprawling to 1 m across.

Baloskion tetraphyllum subsp. *tetraphyllum* in cultivation

Many other *Baloskion* species also prefer abundant moisture or even wet feet for some of the year, though they are all reasonably drought-tolerant once established. These are smaller, almost leafless plants, few of which are particularly interesting to gardeners though *B. pallens* in the form of the narrowly upright, 1 m-high cultivar 'Didgery Sticks' and the flat-stemmed Flat cordrush (*B. complanatum*) which is only half this size are sometimes available. Mountain cordrush (*B. australe*) is the most cold-tolerant species, and is found from the alpine regions of New South Wales to Tasmania. The slender, 70 cm stems tipped with reddish flowerheads can be strikingly ornamental, and stand out against the lime-green sphagnum bogs where the plant is often found. As with most other cordrush groups, you need both male and female plants of any *Baloskion* species to produce viable seed. Sown fresh, it is not hard to raise on moist, peaty soil, and smaller plants can be divided easily.

Garden specimen of *Baloskion tetraphyllum* subsp. *tetraphyllum*

Bambusa (Poaceae)

There are four (possibly three) native bamboo species, only two of which have been cultivated outside the tropics; of these, the *Schizostachyum* species from Murray Island is discussed separately later. Floodplain bamboo (*B. arnhemica*) is the largest-stemmed native species and is often the dominant plant along streams in the Northern Territory. A clumping plant ~10–12 m tall and several metres wide, it is regarded as a single species at present though there are noticeably distinct populations with different flowering patterns and seasons.

Bambusa arnhemica

This is a small-leaved species with distinctive lime-green culms. It needs constant soil moisture and is sensitive to cold. Reasonably fresh seed is not hard to germinate on moist soil in warm conditions, and smaller seedlings are not hard to transplant. The poles cut from this plant are reputed to be very strong.

The unusual scrambling bamboo *Muellerochloa moreheadiana* from tropical Queensland, and the slender-caned, large-leaved, clumping *Neololeba atra* found from northern Queensland to New Guinea and the Philippines, have rarely been cultivated outside the tropics.

Baumea (Cyperaceae)

Twigrushes are found across much of southern Australia, with some occurring as far north as New Caledonia. Most will grow permanently in shallow water and can spread over considerable distances with time, but they are not hard to control when confined to a container. They are more usually found in places that dry out for at least a few months in summer. In ideal conditions twigrushes may cover an entire wetland and, depending on their size, provide significant habitat for a range of aquatic animals from dragonflies to frogs and smaller fishes. Many types of waterbirds from ducks to grebes and ibises make floating nests from twigrushes, anchored by the roots but able to rise and sink with changes in water level.

Jointed twigrush (*B. articulata*) is the most dramatic and handsome species with its upright 2 m bottle-green stems and hanging tassels of tawny-brown flowers to 30 cm. Backlit by bright sunlight the stems look jointed like a compressed bamboo cane, with neatly spaced internal partitions only 1 cm or so apart. This evergreen plant looks best when permanently flooded, but it will still look good even if its roots dry out every summer. The strappy-leaved Preiss's twigrush (*B. preissii*) from Western Australia is a more slender and slightly smaller relative with weeping stems. It grows in rich, permanently wet soils or in shallow water.

Soft twigrushes (*B. arthrophylla* and *B. rubiginosa*) are still more slender and reach only 1 m or so, but are particularly fine when in flower. The flowers of the even slenderer Bare twigrush (*B. juncea*) are barely worth commenting on, but the combination of its very upright carriage to ~70 cm and the densely packed blue-grey stems make it an excellent sculptural plant throughout the year. This is also the most salt-tolerant species and it is not uncommon in coastal lagoons around much of the eastern and western coastlines of Australia. Square twigrush (*B. tetragona*) is sometimes available in nurseries. It is the least spreading of the species, often growing into a tidy mound around 1 m high and across. Not particularly different from other twigrushes at a distance, the deep-green, four-sided stems need to be looked at closely to fully appreciate their elegant structure.

Baumea arthrophylla

Baumea articulata

Baumea juncea

Baumea preissii

Baumea rubiginosa

Bolboschoenus (Cyperaceae)

Clubrushes are a diverse group formerly lumped together as *Scirpus*: see also the more distantly related *Isolepis*, *Ficinia*, *Schoenoplectus* and of course the one remaining true native *Scirpus*, described later. *Bolboschoenus* are the most reed-like of the clubrushes, with slender leaves peeling outwards from triangular stems. These sedges will grow in shallow water year-round and may spread over considerable distances with time. In drier conditions they become dormant, dying down to rough-skinned, edible tubers around late summer and autumn.

Bolboschoenus caldwellii

When fully ripe, their cocoa-coloured flowerheads hang in clusters like miniature pine-cones at the tip of the stalk. Marine clubrush (*B. caldwellii*) tolerates some salinity and is

Bolboschoenus fluitans

perhaps the most abundant species, growing to 1.2 m at most. It can be recognised by the way the leaves are held more-or-less upright, though their tips may droop. The other two species are usually much larger, and River bamboo (*B. fluviatilis*) with pendulous, strap-like leaves up to 11 mm wide is the tallest at up to 2 m. Marsh clubrush (*B. medianus*) is a little smaller and slenderer but is otherwise similar. All three produce edible tubers which are best harvested while young and red-skinned: as they mature they become darker and woodier, and their taste becomes increasingly displeasing!

Bothriochloa (Poaceae)

Redgrass (*B. macra*) is a fairly coarse-looking, reddish-green grass that spreads slowly by runners; its autumn seedheads look a little like some awnless wheat varieties. It has mainly been grown as a rough, drought-tolerant lawn that doesn't get too tall even left unmown, but is not particularly frost-tolerant so is mostly grown in more northern areas.

Bulbine (Asphodelaceae)

Bulbine lily (*B. bulbosa*) is also known by many less flattering names reflecting its onion-like appearance, including Wild onion, Leek lily and Yellow onion weed. It is found across much of eastern Australia from the subtropics to Tasmania, and well up into subalpine areas. On a recent trip to the Grampians (Gariwerd) National Park in south-western Victoria Jan and I were struck by the glowing beauty of its flowers against sheer cliffs, beneath a lowering grey sky. I crawled as close as I dared to take the photo shown in this book, and on turning back found myself looking into the eyes of a wallaby 3 m away, browsing on another patch of Bulbine at the crumbling edge of a vertiginous 300 m drop – it obviously had a better head for heights than I do.

Bulbine does well in a wide range of growing conditions from open grassland to eucalypt forests, flowering through spring and summer in constantly moist soils, though reasonable drainage is essential. The fleshy, grey-green leaves form tight tussocks to ~75 cm high, with taller stalks carrying a pyramid of yellow, starry, fragrant flowers through spring and summer, each individual flower opening for just one day. In drier conditions plants will die back to their sweet, fleshy corms, which can be eaten after roasting. This species is easily grown from seed stored dry for several months, or by separating individual corms.

Bulbine bulbosa growing high on a cliff face

Capillipedium (Poaceae)

Summer-flowering Scented-top grasses or Spicy-top grasses (*C. spicigerum* and the similar *C. parviflorum*) form upright tussocks to 1.5 m high when in flower, and are topped with delicate, open clusters of spikelets. Despite their common name they are not particularly fragrant unless the flowerheads are crushed to release their fruity odour. These grasses are mostly tropical to subtropical, extending as far south as Sydney. They are grown and harvested on a commercial scale in Queensland.

Carex (Cyperaceae)

The most grass-like of native sedges in general appearance, many of these plants form tussocks with relatively narrow leaves which may be sharp-edged and capable of causing unpleasant cuts if you insist on stroking them, while others will run so that a single plant can cover large areas with time. The flowering stems are usually triangular with spikelets arranged in dangling or upright cylinders, the slenderer male flowerheads always above the female. Many of these sedges are found near water, and the most widespread is the aquatic Tassel sedge (*C. fascicularis*) which forms a bright-green mound 1 m high and wide in good conditions. It is found from New Guinea to Tasmania and is very shade-tolerant, growing in water up to 15 cm deep as well as on damp soils.

Tall sedge (*C. appressa*) and the closely related Basket sedge (*C. tereticaulis*) also grow in seasonally flooded places but are much more tolerant of drought and will thrive even in a normal garden bed as long as it never dries out completely. The tussocks of Tall sedge reach well over 1 m and the sharply triangular, poker-like flowerstalks spread a little wider. Basket sedge tussocks are generally a bit smaller, although some clones grow much larger. In the wild, the two species sometimes hybridise. Other smaller *Carex* species including *C. breviculmis* are woodland plants, while *C. tasmanica* with curled leaf tips thrives even on heavy clays which may be flooded for months and grow as small tussocks to only 30 cm or so in height.

The widespread Lizardtail sedge (*C. gaudichaudiana*) reaches up to 1 m in height and spreads by runners to cover large areas in flood-prone places, and though the small spikes are attractive and resemble the scales of a black-and-green lizard tail some forms rarely flower, and seed is even less commonly found. *C. bichenoviana* is another running species though only half the height and is equally widespread though usually in higher and drier places, with a blunter and broader flowerhead sometimes

Carex appressa

Carex fascicularis

Carex gaudichaudiana

Carex tasmanica

resembling a mace. The *Carex* species that regularly form flowerheads are almost invariably easy to raise from seed, and all of them can be readily propagated by division.

The so-called blue form of *C. gaudichaudiana* is actually the European *C. flacca*, introduced from Bruny Island in Tasmania where it is a well-entrenched weed. It has also been sold as *C. glauca* and *C. gunniana* 'Bruny Island' form.

Carex tereticaulis

Caustis (Cyperaceae)

Curly-wig and its relatives are among the most frustrating grass-like plants for gardeners, as they include some of the most spectacular native dryland sedges but are difficult to establish. Like many native plants from eucalypt woodlands they don't like much nitrogen or phosphate in their soil, don't reliably produce viable seed and are slow-growing even in the wild. Some of these sedges have been unsustainably harvested as cheap packing with flower exports, which are used in dry flower arrangements overseas. Harvesting from the wild appears to have been banned in most places but continues in many inland areas.

Curly-wig (*C. flexuosa*) is found from Victoria to Queensland. It is far from the most spectacular member of this group, as the subtropical *C. recurvata* is even more elaborately twisted and whorled. Mature plants of these species tend to sprawl as they grow, and can reach ~1.5 m in height and width, though the top growth is often sparse. The poorly named Koala fern (*C. blakei*) has a similar range but is more like a rosette of low-growing, pendulous foxtails up to 1 m in width – definitely not koala-like! There are also several interesting Western Australian species, though nothing seems to be known about their cultivation.

Caustis blakei

Caustis recurvata

It is possible that for good growth these plants need a specific mycorrhizal fungus which is only found in bush soils, as I have had reasonable results transplanting Curly-wig seedlings from a graded gravel roadside (where they would have been scraped off by a grader as part of routine maintenance) two years after a bushfire swept through. This also suggests that potassium in the form of wood ash may be useful in their soil, and perhaps help to trigger seed formation in mature plants.

Cenchrus (Poaceae)

Swamp foxtail grass (*C. purpurascens*) forms handsome tussocks with purplish flowerheads on upright stems to 1.5 m high, and grows in moist to wet but well-drained places from tropical Queensland to southern New South Wales. Among grasses it should be a must-have for any native garden in eastern Australia, suiting almost any situation or style from wild pond fringes to strictly formal borders. There has been much confusion between this species and the non-native and very weedy Fountain grass (*C. setaceum*). Both were widely sold under the name *Pennisetum alopecuroides* in the early 2000s, along with occasional other introduced grasses from the same genus.

There were other confusing factors. Not all forms of Swamp foxtail grass sold in nurseries even now are grown from indigenous seed, nor is it easy to decide which plants of this species out there in the wild *are* indigenous in the sense that they were already here before the arrival of

Cenchrus purpurascens

Europeans. Genetic testing shows that some strains of Swamp foxtail grass growing around desert springs probably arrived from South-East Asia hundreds of thousands of years ago, and it is likely that other strains in warmer parts of eastern Australia have a comparably long history.

The photo should give you a good idea of what non-weedy, indigenous Swamp foxtails look like, including their relatively compact bottlebrush flowerheads as opposed to the usually more elongate and tapered heads of many imposters. New plants can be raised from seed or by division. They grow best in soils that are never completely dry, exposed to full sunlight. Once established they will tolerate serious frosts, and fertilisation should be minimal as it is more likely to encourage weeds than improve the looks of this lovely plant.

Centrolepis (Centrolepidaceae)

These miniature tussocks are so rarely noticed that they don't even have a common name, yet on seasonally wet soils (especially on heavy clays) they may grow so densely as to form a zero-maintenance lawn just a few centimetres high. Their flowerheads are tiny, with horned or hairy structures carried above the leaves, and fascinating in their intricate detail if you are prepared to get down and dirty looking at them. All species are easily propagated by teasing tussocks apart and will self-seed once established, but need a reliable supply of water for the first year or two, and may be shaded out by taller plants. The dark-green hemispherical tussocks of *C. fascicularis* are topped by tiny, feathery, white pennants. The flatter and more sprawling *C. glabra* is also attractive but has rarely been available through nurseries.

For lovers of aquarium plants, this is the most abstruse story you will read in this book, included for its all-round improbability. *C. drummondiana* is a south-western Australian species (formerly placed in *Eriocaulon*, or sometimes *Trithuria*) that normally grows on dry land like all its other relatives, but it has been co-opted into the aquarium trade despite a few problems. To succeed with this species in an aquarium, it needs at least 100 μmols of light (a relatively obscure unit of light intensity that won't mean much unless you are an aquarium plant collector and already own a wide array of specialised monitors for light levels and water conditions), and as it only grows a few centimetres high like most other *Centrolepis* it won't do well under taller plants. You also

Centrolepis fascicularis

need to make sure additional carbon dioxide is delivered downwards to the aquarium floor where it must be planted.

On top of everything else the dubious trade name is 'Blood Vomit'. I have been intrigued enough by this plant to start experimenting with other native *Centrolepis* as aquarium plants, but for less ambitious readers with a similar interest in these tidy, compact miniature grass-like plants my advice is to just plant them in silty, nutrient-poor soils in deep pots, and water them once or twice a week in dry weather. They'll grow faster than under water and also reward you with fine shows of brave little floral banners!

Chaetanthus (Restionaceae)

Western Australian cordrushes from seasonally wet heathlands, these are readily grown from seed sown onto a sandy, peaty soil mix, as they regenerate naturally this way after fire. They can also be grown by division in spring, just as the plant starts to put out new growth. The only species readily available at the present time is *C. aristatus*, a slender-leaved, open tussock to 80 cm in height and a little wider when in flower, with masses of pendulous flowers on male plants, rather like a smaller and more delicate-looking *Leptocarpus*.

Chaetanthus aristatus male flowerheads

Chloris (Poaceae)

Windmill grass (*C. truncata*) is one of the most common and widely distributed native grasses, particularly in drier inland areas where it

Chloris truncata

competes effectively even against invasive introduced grasses. The drought-tolerant tussocks are topped with radially spreading ribs forming an umbrella-like frame, with larger plants reaching 50 cm in height and 1 m in width. Strangely, this common and easily propagated grass (the seed is freely produced and easily germinated) is rarely available through nurseries, presumably because it is relatively short-lived. However, with its distinctive habit and low maintenance requirements, plus a willingness to self-seed even among weeds, it deserves to be more widely grown, especially in arid areas.

Chorizandra (Cyperaceae)

Bristlerushes are a small group of distinctive and attractive sedges which will grow in permanently flooded, shallow water yet survive months or even years of drought once established. The flowerheads are set into the upper part of the stem, sometimes with a broad, flaring bract behind them, creating an effect resembling a hood. They include Heron bristlerush (*C. cymbaria*) and the more rounded Globular bristlerush (*C. sphaerocephala*), both of which reach ~1 m in height and spread indefinitely though slowly.

Chorizandra sphaerocephala

The hoods behind the flowerheads of the smaller Black bristlerush (*C. enodis*) are less obvious, so from some angles they resemble a more delicate-looking version of *Ficinia* and the newly opened flowerheads look like miniature

Chorizandra enodis in habitat with flowerheads inset

echidnas climbing the stems. This is perhaps the most attractive and drought-tolerant species, with slender, wiry, grey-blue stems to ~60 cm, and it will continue to look good even after a dry summer. All bristlerushes are easily propagated by division but can be slow to re-establish, and are not hard to grow from seed if you can find any.

Cladium (Cyperaceae)

Leafy twigrush (*C. procerum*) is among the larger and more spectacular native sedges with broad, deep-green, sharp-edged leaves forming open

Cladium procerum

tussocks ~2 m wide, while the large, globular masses of flowers carried on 2.5 m stems change from golden-brown to a rich brown as they age, initially resembling a cauliflower in size and shape. This giant usually grows in seasonally flooded soils near the coast but doesn't thrive when permanently waterlogged. Seed is not reliably set, but Leafy twigrush is easily multiplied by splitting the congested groups of young plants which form along the nodes of the tall flowerspikes.

Conostylis (Haemodoraceae)

A group of Western Australian tussocks related to kangaroo paws, these plants form tidy, low clumps of grass-like or rush-like foliage which may be almost hidden under masses of (usually) yellow flowers in spring. Only a few of the 40 or so species are in cultivation. The best known is Grey cottonhead (*C. candicans*) which is named for its silver-grey leaves and the cottony, tan-coloured seedballs which follow the flowers. Grey cottonhead reaches ~40 cm in flower, and sprawls a little wider as the seedheads mature.

Other species with silver-furred leaves such as the white-flowered White cottonhead (*C. setosa*) and golden-flowered Bristly cottonhead (*C. setigera*) are only around half that size in both height and spread.

Conostylis candicans

C. juncea has particularly slender leaves that look a little like some Juncus, until masses of clustered yellow flowers appear in spring, while Prickly conostylis (*C. aculeata*) is stiffer-leaved and more densely packed. All species need good drainage and full sunlight, and most are drought-tolerant once established. Propagation from seed does not seem to be reliable, and divisions need protection from hot sun until their roots are well-established.

Cordyline (Asparagaceae)

Slender palm-lily (*C. stricta*) is an elegant, bushy plant generally found from north-eastern New South Wales to southern Queensland but it will grow as far south as Hobart and some fine specimens can be seen in older, well-established gardens. It is included here as it is sometimes mistaken for a bamboo, but the 50 cm strap leaves growing directly from crowded nodes along the branching stems are very different from any of the true grasses, and the antler-like flowerheads develop glossy, purple-black berries in late summer.

Thriving in both sun and heavily shaded conditions in frost-free places, it can reach 5 m in height and mature clumps can be almost as

Cordyline stricta with flowers and fruits inset

wide. It can also be grown as a moderately drought-tolerant pot plant which reaches 2–3 m, shedding its lower leaves more readily to expose neatly spaced scars along the canes. Propagation is usually from cuttings taken in the warmer months, but it can also be grown from seed started in a greenhouse. The similar but smaller orange-berried *C. congesta* with scalloped leaf bases goes under the same common name but is not often available as it seems more sensitive to cold.

Crinum (Liliaceae)

There are several native species of these showy lilies, of which the broad-leaved Swamp lily (*C. pedunculatum*, sometimes referred to *C. asiaticum* var. *pedunculatum*) is sometimes included among grasses in nurseries. It doesn't really look any more grass-like than most other lilies, and with time will raise itself up on a fleshy trunk to 1 m high so the whole plant may be 2 m tall and almost as wide, looking something like a succulent grasstree! The plants take a long time to reach this stage, and they are usually grown for their clustered heads of large, tubular, lily-like flowers which appear regularly long before any sign of trunk development. Despite the common name this plant won't survive more

Crinum pedunculatum in habitat at Myall Lakes, New South Wales

than a few days of waterlogging. It prefers rich, moist soils in a frost-free location and it flowers even in surprisingly dense shade. It is easily grown from fresh seed sown into warm soil, or with a little more effort can be divided to produce a faster-maturing plant.

Curculigo (Hypoxidaceae)

Palm grass (*C. capitulata*) has also been known as *Molineria recurcata* and combinations of these names have been shuffled back and forth for some time, with hopefully a consensus settling on the name used here. Whatever species name you choose to use, these are handsome, palm-like plants with strongly pleated leaves, reaching around 1 m or more in height and needing moist, rainforest conditions and protection from hot

Curculigo capitulata

sun for best appearance. Found through much of tropical Asia, it extends naturally into northern Queensland but is also happy in subtropical areas and will tolerate surprisingly cold conditions though not frost.

Cycnogeton (Juncaginaceae)

Water ribbons are a distinctively Australian group of eight aquatic species previously included in *Triglochin*, and may be the dominant plants over large areas of wetlands. With their ornamental strap-like leaves and dramatic, often upright, green flowering columns, they vary mainly in size and degree of adaptation to drought. The tubers formed among the roots are good to eat raw or cooked (though too small to be worth harvesting in some species), while seeds and leaves are eaten by various waterbirds. The leaves are a useful nesting material for swans and grebes. The shapes and arrangements of the rounded clusters of seeds are useful identification characters.

Common water ribbon (*C. procerum*) is found across much of south-eastern Australia in waters which may or may not dry out each year, and at up to 1 m is one of the larger and broader-leaved species. It is not bothered by drought and is so adapted to dry conditions that it will put

Cycnogeton alcockiae flowering

out leaves when rain falls after a dry spell, then just sit there replenishing its tubers unless more rain falls to flood it, triggering flowering. Many-fruited water ribbon (*C. multifructum*) is even more showy, with upright flowerstems covered with up to 1,000 closely packed fruits, an attractivc sculptural contrast to the strap leaves floating flat on the water.

The smaller species are well-adapted to pools that often dry out months after seasonal rains come to an end: these are Pygmy water ribbon (*C. alcockiae*) in south-eastern

Cycnogeton multifructum fruiting

Australia, and Narrowleaf water ribbon (*C. lineare*) in the south-west. All water ribbon species are easy to grow from seed planted onto waterlogged soil or in shallow, clear water. As various species of attractive water ribbons are found across much of Australia, there is no reason to import species for garden use from outside their natural range.

Cymbopogon (Poaceae)

The best-known grass in this group is the exotic lemongrass (*C. citratus*) used as a flavouring in Asian cookery, but there are several related Australian species which are moderately fragrant and also much more interesting to look at than the bland, lime-green tussocks of the condiment plant. Barbed-wire grass (*C. refractus*) is the most familiar species, reaching up to 1 m high with arching flowerspikes half as long again, looking remarkably like barbed wire though they are fortunately soft to the touch. These shift from blue-green to bronze-brown as they mature, but look untidy beyond this stage and should be removed before they set seed, to encourage flowering in the next season.

Even more drought-tolerant is Native lemon grass (*C. ambiguus*), which is more strongly fragrant and can be twice as tall, while *C. gratus* (which appears to have no common name) is an upright Queensland species ~40 cm high that looks drought-stricken much of the time, but a close look reveals healthy green growth beneath the dusty-looking surface. Although Barbed-wire grass is found as far south as Victoria these are essentially plants of warmer and often drier places ranging up into the tropics. They will not

Cymbopogon gratus

Cymbopogon refractus

do well in prolonged cold, wet weather or with anything more than the occasional light frost. All species are easily divided to give new plants.

Cynodon (Poaceae)

The common and widespread Couch or Bermuda grass (*C. dactylon* var. *dactylon*) is an introduced weed, and although there may be some less weedy, indigenous forms (var. *pulchellus*) of this species these have apparently never been available through nurseries.

Cyperus (Cyperaceae)

The flatsedges include some of the most striking native sedges, many of which are ecologically important as habitat for a wide range of wetland animals. They form large stands in and around wetlands and along seasonal streams, and hold the shoreline together during floods. Among the largest and most attractive forms is Tall flatsedge (*C. exaltatus*), which can be abundant along parts of the Murray River and towards the coast from around Sydney northwards. This upright and fairly narrow-leaved tussock can become very bushy and carries densely packed masses of golden-brown spikes, reaching ~1.5 m high.

The more southern Leafy flatsedge (*C. lucidus*) is even taller in good conditions, with large, well-opened heads of metallic bronze to 2 m above broad, deep-green, razor-edged leaves. This species prefers more shaded localities where it will spread slowly over long distances, but it will also grow vigorously in full sun if kept permanently wet. Wiry flatsedge (*C. gunnii*) is a slenderer relative which will grow in deeply shaded conditions, forming a low mound around 1 m high, though with its much narrower, leathery leaves it also does well in hotter, inland areas and survives serious drought relatively unscathed.

Several other smaller and narrower leafrushes are still more drought-tolerant once established; for example, the slender-leaved *C. gracilis* which will only thrive in dry woodlands or on rocky outcrops. Others, such as the edible-tubered Downs nutgrass (*C. bifax*), Nalgoo (*C. bulbosus*) and Yelka (*C. victoriensis*), are found on inland floodplains which may dry out completely for many years at a time, coming to life only during wetter years. Few of these reach more than 1 m in height, and they are often smaller in drier seasons.

An interesting natural pairing of drought-adapted sedges is Native umbrella-sedge (*C. vaginatus*) which has elegant, slender and leafless flowerstems crowned with a circle of leaves around the golden-brown flowerheads, forming rounded mounds to 1 m high. Inland umbrella-sedge (*C. gymnocaulos*) is only around two-thirds that size and has almost no leafy foliage on top – presumably an adaptation to longer periods of drought. These supposedly separate species overlap in range and there are intermediate plants which can't be tidily fitted into either species, though the most strikingly ornamental plants of this group are usually found closer to

Cyperus bifax

coastal areas. Both will grow in permanently flooded conditions and thrive permanently in reasonably moist soils. It is only their degree of adaptation to drought which makes them look so different at the extremes of their blended range.

Cyperus exaltatus

Cyperus gunnii

Some short-lived or even annual flatsedges such as *C. polystachyos* are also worth growing, self-seeding to form scattered stands of fresh green, upright tussocks with flattened, open flowerheads. Nutgrass (*C. platystylis*) is a handsome, upright species with particularly nice, symmetrical, golden flowerheads. It is found scattered across much of northern Australia though it has also naturalised around Sydney, so don't plant it near wetlands outside its natural range. These shorter-lived species only reach 1 m even in the richest soils. Most flatsedges are easily grown from seed sown onto moist or

Cyperus gymnocaulos

Cyperus lucidus

Cyperus platystylis

wet soils as well as by division, the main exception being Leafy flatsedge which may not produce much seed due to a fungus that attacks the developing seed; fortunately, this doesn't affect the attractiveness of the flowerheads.

Cyperus vaginatus

Deschampsia (Poaceae)

Tufted hairgrass (*D. cespitosa*, not *D. caespitosa* as it is often spelt) is an attractive tussock grass found in alpine areas in Australia; it is also widespread in the northern hemisphere. The low tussocks are only

~50 cm wide but produce shimmering, open flowerheads to 1.5 m high which look particularly striking when wetted by dew, which often occurs in their high range. Some botanists have suggested this is an introduced species because Australian populations are so far removed from the rest of the species' known range, but as an uncommon, non-weedy grass found patchily at high elevations far away from where any introductions are most likely to have been made, it is now generally accepted as a native plant. Seed doesn't seem to be formed readily in warmer conditions, but it is easily propagated by division.

Desmocladus (Restionaceae)

These curious, often twisted plants from sandy heathlands in Western Australia can form dense, self-supporting, slowly spreading tangles on open ground and are almost drought-proof once established. They usually set abundant seed which germinates after fire, but the only species regularly available through nurseries in eastern Australia is the 40 cm *D. flexuosus*, readily propagated by division in spring.

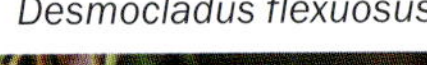

Desmocladus flexuosus

Dianella (Liliaceae)

With masses of small, often vivid, blue flowers followed by even more showy blue or purple berries, and agapanthus-like strap leaves, flax-lilies aren't all that grass-like though they are usually stocked in the grasses section of nurseries. There are dozens of species which vary in habit and height, but only a few are widely grown. Tasman flax-lily (*D. tasmanica*) is the largest and most striking of these, reaching up to 1 m and carrying generous bunches of fat, luminous, blue berries above. This is one of the most cold-tolerant species and two smaller selected forms (~40–50 cm) are also widely available. 'Tasred' shows a reddish tint at the base of the foliage in colder weather, while 'Blaze' has purple foliage which deepens in colder conditions and is reputed to be heat-tolerant, though it needs some protection from the sun from Sydney northwards. 'Cherry Red' has particularly fine purplish-red leaf bases that contrast pleasingly with the fresh green foliage above, while 'Variegata' has handsome white stripes along the leaf edges, though its flowers are pallid and insignificant.

The sky-blue flowers of Blue flax-lily (*D. caerulea*) are paler than those of many other species. Several named forms are available, from 'Curly Tops' with particularly leafy, upright flowerstems to

Dianella caerulea subsp. *asella* 'Curly Tops'

Dianella tasmanica berries

80 cm, to selected smaller forms rarely reaching above 50 cm such as 'Cassa Blue' with bluish foliage and 'Little Jess' forming compact clumps of ~40 × 40 cm. Pale flax-lily (*D. longifolia*) is not particularly distinctive except in its white-flowered form, and Spreading flax-lily (*D. revoluta*) is available in several small-growing forms. All flax-lilies do best with protection from hot sun, and need reasonably moist soils or they won't flower. Propagation is easy from divisions though it may be some time before new roots appear, and seed which has been separated from the fleshy fruit and kept dry for a few months is not difficult to germinate.

Dichanthium sericeum

Dichanthium (Poaceae)

Slender-stalked and very upright to around 1 m in height, Silky bluegrass or Queensland bluegrass (*D. sericeum*) is found across much of the mainland as far as southern Victoria, and is one of the most readily available native grasses for good reason. The pale, dusty-looking, pastel-blue foliage and fuzzy forked flowerheads held up to 1 m high on stiffly upright stems are decorative through most of the warmer months, and established plants are reasonably drought-tolerant once established. Easily propagated from seed or by division, it needs abundant sun to grow well and produce best foliage colour.

Dichelachne (Poaceae)

Long-hair plumegrass (*D. crinita*) is the most widespread of this group, found across all of Australia apart from the tropics or in deserts and thriving even close to the coast. It is also the most cultivated species. It has long, tight, silvery plumes standing up to 1 m tall from a small tussock, and is an elegant and distinctive grass which looks particularly good massed under eucalypts. Shorthair plumegrass (*D. micrantha*) is similar but has a more compact, slender flowerhead. Easily grown from seed, these fast-growing plants can also be propagated by division.

Dichelachne crinita

Dichopogon (Asparagaceae)

Chocolate lily (*D. strictus*, formerly *Arthropodium strictum*) is closely related to Vanilla lily (*A. milleflorum*) and is comparably cold-tolerant as well as thriving on most soils. The chocolate-scented purplish flowers are held well above the foliage in spring and look like the distantly related fringe lilies (*Thysanotus*), though without tassels. The elongated tubers can be eaten raw or lightly roasted, but become bitter if left too long after harvesting.

Dichopogon strictus

Dietes (Iridaceae)

It seems ridiculous to include a plant known as Wedding iris (*D. robinsoniana*) in a book on Australian grasses, and it is here only because it has already appeared in other people's books and articles in this improbable guise. Broad-leaved with large, white, yellow-centred flowers, with leaves reaching 1.5 m though the plant itself is usually narrower, it is strikingly similar to the true irises in appearance. It qualifies as a 'native' only on the grounds that Lord Howe Island is an Australian territory. Unlike many other plants in this book it requires both good drainage and constantly moist soils with

Dietes robinsoniana

plenty of sun, and is easily damaged by frost, but on the plus side grows well near the sea.

Diplarrena (Iridaceae)

Butterfly flag (*D. moraea*, also known as White Iris) is a close relative of the Wedding iris (previous entry), and the two could be confused when in bud or when the papery, three-celled seedpods are formed, although *D. moraea* is narrower-leaved and more stiffly upright. The yellow- and purple-centred white flowers appear from spring to early summer, and don't last all that long though the fascinating papery seed capsules remain attractive well into autumn. A larger, alpine version of this species from Tasmania known as 'Amethyst Fairy' is sometimes regarded as a separate species (*D. latifolia*).

Butterfly flag is not fussy about soils as long as they are free-draining, though the larger-growing Tasmanian forms found in formerly glaciated areas of Tasmania suggest that neutral, mineral-rich soils promote growth. This species does well in pots and doesn't need repotting often if lightly topped up with a low-nitrogen fertiliser suitable for most native plants, towards the end of winter. Easily grown from seed or by division, it does best in full sun, but does not like prolonged hot, dry weather.

Distichlis (Poaceae)

Australian saltgrass (*D. distichophylla*) forms low, running mats of slender, grassy fans on wet soils and clays, tolerating both salt and once

Distichlis distichophylla

established a fair degree of drought, so it has considerable potential as a rough lawn grass in some difficult situations. The flowerheads are relatively small, pale-green spikes which makes seed collection difficult, but even small divisions without roots will usually strike easily, even when watered only with fresh water.

Doryanthes (Doryanthaceae)

The two species of giant lily form impressive, broad-leaved clumps with dramatic flower stems of brilliant red or pink flowers in spring, sometimes also showing cream or white. Each rosette within a mature clump flowers only once and may take up to 13 years to reach that stage, dying soon after seed is set though the main clump usually remains green. Neither species is hard to grow from reasonably fresh seed planted into moderately rich, preferably deep soil and both will grow in sun or semi-shade, benefiting from some watering and moderate fertilising with blood-and-bone. Selected colour forms are propagated by division of the underground stem, most easily done while the plants are relatively small.

Giant spear lily (*D. palmeri*) is from the eastern New South Wales–Queensland border but will grow well in frost-free coastal regions as far south as Hobart. It is by far the larger of the two at well over 2 m high and often twice as wide, with even longer flower stems. These start off looking like a stalk of asparagus, then droop as they become longer until they are suspended not far from the ground. Each stem carries hundreds of large buds opening into very attractive glossy flowers, followed by thick, fleshy pods dangling along much of the stem, looking like a surrealist's dream of a bunch of red bananas.

Doryanthes palmeri

Gymea or Flame lily (*D. excelsa*) is a more southerly plant from the central New South Wales coast and will tolerate some frost, but extreme cold will kill any seed that is forming. Clumps of this softer-leaved species are 'only' around 1 m high, and look spectacular when in flower among the eucalypts and rock outcrops of their natural habitat. The flower columns are vertical, up to 4 m tall and are topped with a 30 cm ball of (usually) red flowers.

A multi-headed plant of *Doryanthes excelsa*

Dracophyllum (Ericaceae)

This genus of remarkable 'dragon-leaf' heaths is scattered from Tasmania to New Caledonia, although most of them are New Zealand endemics and only a few could reasonably be considered grass-like. For the many gardeners who would love to grow a Giant grass tree (*Richea pandanifolia*, from the same family) but are reluctant to move to Tasmania to satisfy this perfectly natural desire, there are two related species growing in and around the Sydney area that don't need a cold climate to thrive.

D. secundum (neither species seems to have a common name, so let's just call then Dracophyllums) is a small shrub to ~1 m high, found on sandstone from the coast inland to the Blue Mountains. Clusters of typical heath flowers appear at the top of each stem in winter or spring and are attractive though subdued in colour. *D. oceanicum* is a taller species to ~2 m, found only around the Jervis Bay area and forming sprawling clumps in moderately shaded gullies near the sea. Both species need moist conditions at all times.

The problem with these two species is that they are unfamiliar and tricky to establish, so you won't find them in any garden centre. Seed isn't reliably produced and I don't know any account of one being successfully raised from that source. Like most heaths they can be grown from spring

Dracophyllum oceanicum

cuttings with a bit of care (and rooting hormones won't hurt, either) if kept in shaded conditions, though I have not been able to keep them alive through winter as our climate in Victoria's Otway Ranges is more like southern Tasmania than coastal New South Wales.

Echinochloa (Poaceae)

Most swamp millets and barnyard grasses are introduced weeds, but the subtropical Swamp barnyard grass (*E. telmatophila*) with its furry, long-awned heads on arching 1.5 m long stems is an attractive native which

Echinochloa telmatophila

grows in and around streams. An annual, it must be grown from seed, but will naturalise readily around ponds or on seasonally wet ground.

Eleocharis (Cyperaceae)

Spikerushes are mostly small to medium-sized, running sedges with more-or-less upright stems, tipped with a pointed flowerspike whose proportions resemble the head of a match. The larger species include many native variations of Chinese water chestnut (*E. dulcis*) with hollow, cylindrical stems to 2 m high, growing mainly in the tropics though these also do well in warmer areas further south. As with many other plants that spill into Australia from South-East Asia, it is likely that this species is genetically variable. There may be native forms that

Eleocharis acuta flowering

Eleocharis equisetina

Eleocharis geniculata

Eleocharis sphacelata

produce better-quality tubers and are at least as productive as the commercial variety most often grown.

Tall spikerush (*E. sphacelata*) is much more widespread, reaching Tasmania and even New Zealand. This vigorous plant will grow in deeper waters to 3 m, and although it is handsome it is by far too large for garden ponds. However, it makes an excellent shelter and nesting habitat for many waterbirds, as do the twigrushes (*Baumea*). As it can take over smaller dams even in waters up to 3 m deep, I recommend less vigorous aquatic sedges for habitat plantings, such as Jointed twigrush (*Baumea articulata*).

The smaller species include the widespread Common spikerush (*E. acuta*) to 1 m high though usually lower in seasonally dry places. It is a much denser-growing plant which will tolerate drying out once established. There are also various smaller relatives which are not always easy to identify and are rarely cultivated. Some of the smallest can be used as attractive zero-maintenance lawns for seasonally wet places, as long as potentially competing weeds are removed before planting as the smaller species only reach a few centimetres in height; *E. pusilla* is perhaps the most delicate-looking and ornamental of them.

One of the most elegant upright species is *E. equisetina* from warmer regions to the tropics, with a particularly long flowerhead, and the even more tropical *E. geniculata* with unusual, rounded heads looking a little like beads, quite different from more typical species of the group. *E. geniculata* is often deeply submerged during the wet season, and thrives as a true aquatic in a clear, warm pond. All spikerushes are easily propagated from seed (if you have the patience to collect it) but are more quickly established from divisions.

Empodisma (Restionaceae)

Twining roperushes are slender-stemmed, fairly upright plants which may scramble through other vegetation around wetlands or form slow-spreading, self-supporting mounds up to 1 m high, usually in seasonally wet soils; they also tolerate some waterlogging. Both species will grow in well-shaded situations though seed may not be produced if light levels are too low, and are moderately drought-tolerant once established. Fresh seed sown onto wet soil or even clay will germinate fairly readily. There are two species, of which *E. minus* from eastern Australia is the only one reliably available. *E. gracillimum* from Western Australia is very similar in overall appearance.

Empodisma minus male plant

Eragrostis (Poaceae)

Lovegrasses are mostly nondescript, scrawny tussocks though the plant sometimes described as Lavender grass (*E. elongata*) has been grown for its pale lilac flowerheads and can reach ~80 cm when in flower. More spectacular are the canegrasses, especially the inland species Desert canegrass (*E. australasica*) which is found across much of the warmer and drier parts of the mainland. The elongated canes reach over 2 m from a narrow base and arch dramatically under the weight of the flowerheads,

Eragrostis australasica

while their waxy blue-green colour contrasts dramatically with the strikingly orange soils of the claypans they are often found in. Drought-tolerant in the extreme, this plant may only flower after a rare fall of desert rain, if its roots remain flooded for a few weeks. The smaller Southern canegrass (*E. infecunda*) is not quite as drought-tolerant but is not affected by frost. All lovegrasses and canegrasses can be grown from seed or, more usually, by division.

Eurychorda (Restionaceae)

An upright, slow-spreading Western Australian cordrush sometimes 1 m in height, *E. complanata* has markedly flattened, grey-green stems topped with golden-brown flowerheads. In nature it sets seed freely, but the plants usually available to gardeners have been propagated by division in spring and are probably all one sex. In its heathland habitats this plant thrives with winter moisture, yet it is also drought-tolerant.

Eurychorda complanata

Evandra (Cyperaceae)

The Western Australian sedge *E. aristata* is a strikingly handsome and very distinctive plant when full grown, yet it doesn't seem to have a common name. It forms a large and open tussock to ~1 m high, resembling some smaller sawsedges (*Gahnia*) but with slender and daintily arching flowerstems reaching far beyond the foliage. It prefers reasonably well-drained sites but is able to tolerate lengthy wet periods if it isn't flooded for long. Although rarely cultivated, it is not hard to grow from seed.

Evandra aristata

Ficinia (Cyperaceae)

Ficinia nodosa flowering

Knobby clubrush (*F. nodosa*, incorrectly *Ficinea*) is one of the most widespread, common, widely used and most drought-tolerant sedges. It is often planted in carparks and on nature strips, and arranged in tidy groups in botanical gardens. This is a fine ornamental with upright or sometimes semi-weeping cylindrical stems to ~1 m in height though only half as wide, with globular flowerheads which look particularly attractive when the flowers are first opening. It thrives on coastal sand dunes, around the fringes of wetlands, in slightly saline conditions even in arid areas and in almost any garden soil, and is easily propagated from seed or by division.

Fimbristylis (Cyperaceae)

Fringerushes are slender, upright or semi-weeping sedges of seasonally wet places which tolerate shallow flooding for months at a time. There seem to be dozens of species in warmer parts of Australia, though few are anything to write home about as far as distinctive appearance goes. Most of them form small tussocks less than 30 cm across or high. However, some resemble the smaller spikerushes (*Eleocharis*) and can be used in the same ways, including as potted specimens. In eastern Australia one of the most common species is *F. dichotoma*, but a blue-green form of the more tropical *F. tetragona* is more often available in nurseries. All species are easily propagated by division, or from seed.

Fimbristylis tetragona

Flagellaria (Flagellariaceae)

Whip vine (*F. indica*) is a tropical to subtropical vine very closely related to true grasses. Its leaves are attached directly to the 1 cm-wide climbing stems as in Common reed (*Phragmites*), but the tip of each leaf has a twining tendril to help the vine climb. This is a dramatic and vigorous plant in warm conditions and may need regular removal of unwanted stems, which may scramble their way 15 m up through trees. The flowerheads aren't particularly showy but are pleasingly fragrant – if you can persuade a Whip vine to flower close enough to the ground to

Flagellaria indica

smell them! This species needs protection from the hottest sun as it will not tolerate drying out. It is usually propagated by division of the roots (cut the canes back to less than 50 cm at the same time); vine cuttings may also strike in warmer conditions.

Fuirena umbellata

Fuirena (Cyperaceae)

The semi-aquatic *F. umbellata* looks like a curious, narrow and very upright, five-sided grass rather than a sedge. It will grow in any damp soil or even in shallow, slow-moving water. The small, green clusters of flowers are held above the foliage, and flowering plants are rarely much more than 50 cm in height. Unusual and ornamental, it is short-lived in seasonally dry situations but will self-seed readily whenever conditions become wetter again.

Gahnia (Cyperaceae)

The sawsedges are among the largest native tussocks. They form leafy mounds as high as they are wide, with showy flower plumes held well beyond the sharp-edged foliage which gives them their common name. Although mostly growing in seasonally wet places they are also very drought-tolerant, retaining their good looks and flowerstems even through a dry summer. However, as they burn very readily they should not be planted too close to buildings. In terms of ornamental qualities, sawsedges can be divided into two natural groups: eastern Australian species with bright red or orange seeds, and those with relatively inspicuous seeds.

The best known of the red-seeded group are Tall sawsedge (*G. clarkei*) and Red-fruited sawsedge (*G. sieberiana*), both of which carry masses of shining, blood-red seeds on dark flowerheads and can form tussocks over 2 m high. The flowerheads are even higher, often becoming pendulous as they mature and the weight of the ripening seed drags them down. On Tall sawsedge the feathery, open flowerheads give a cascading effect, while on Red-fruit sawsedge they generally form a tighter and more poker-like column. On a recent visit to Tasmania I fell in love with Brickmaker sedge (*G. grandis*) which is abundant there, though it also grows on the south-eastern mainland. This is a striking species with densely packed, usually very upright flowerstems and orange-brown to reddish-brown seeds.

The other sawsedges are all smaller than these giants, and the smallest of them are rarely cultivated. Cutting or Coast sawsedge (*G. trifida*) is the only large species found in both eastern and western Australia. It forms dense tussocks up to ~1.5 m high and tolerates moderately saline soils. Chaffy sawsedge (*G. filum*) is similar at first glance and is often found with *G. trifida* in eastern Australia in less saline conditions, but it is more tolerant of prolonged flooding. Thatch sawsedge (*G. radula*) is a much smaller and often abundant plant which rarely produces seed, but its masses of black, 1 m high plumes can be very showy among other grasses. The smallest species likely to be seen in cultivation is Rough sawsedge (*G. aspera*), generally no taller or wider than 60 cm and often forming dense colonies in dry, open forests in eastern Australia. Though the seedheads are small and hidden among the fine foliage, the clusters of chocolate seeds have a certain charm of their own.

Gahnia aspera

Gahnia clarkei with seeds inset

Gahnia filum

Gahnia grandis

Gahnia sieberiana

All sawsedges are difficult or impossible to propagate by division or transplanting. They must be grown from seed, and even then they are not always co-operative. For the red-seeded species, the seed needs to be stored for 12–18 months before planting or it won't germinate, and results may still be poor depending on the individual plant or collection region. All the other sawsedges germinate fairly readily (and some will come up like a lawn, so sow them thinly!), with a single exception – Thatch sawsedge rarely produces seed at all. However, if successfully established in a pot, it can usually be divided in spring just as it starts into growth.

Glyceria (Poaceae)

Austral sweetgrass (*G. australis*) is a slender-leaved grass growing in similar conditions to those required by the swamp wallaby-grasses (*Amphibromus*), though it is more abundant in cooler, southern areas. It is also very similar in size and its semi-weeping to upright habit. It is a significant habitat plant, especially for frogs, and will control erosion along streambanks once well established. From a distance the two types of grass aren't easy to tell apart, but the flattened spikes of sweetgrass don't have the prominent awns seen on swamp wallaby-grasses (see the

Glyceria australis

direct comparison of the two in Chapter 1). Propagation is easy by division or from seed.

Gymnoschoenus (Cyperaceae)

Buttongrass (*G. sphaerocephalus*) may form extensive plains on peaty, seasonally wet soils in south-eastern Australia, and are a dominant vegetation type in parts of Tasmania from the coast to high altitudes. Buttongrass thrives best in full sun and relies on fire to prevent competing vegetation in the form of teatrees and paperbarks from shading it out. The low tussocks form a green ball ~60 cm in diameter with long arching stems above, each tipped with a prominent button. Many plants which have been offered under this name by nurseries are misidentified, particularly yelloweyes (*Xyris*) which grow in similar places and even look a little like stunted Buttongrass plants when not in flower. Buttongrass doesn't reliably set seed (particularly on the mainland), although it is not difficult to germinate if you can find it, so it is usually propagated the slow way by division. *Chaetospora* is an old synonym for this species and all plants I have seen sold under this name were misidentified, most of them as *Xyris* species or sometimes Knobby clubrush (*Ficinia nodosa*).

Buttongrass plain in Tasmania

Gymnoschoenus sphaerocephalus at Dove Lake, Tasmania

Gymnoschoenus sphaerocephalus in the Grampians (Gariwerd) National Park, Victoria

Gymnostachys (Araceae)

Settler's flax (*G. anceps*) is a curiosity from subtropical eastern Australia, a narrowly upright plant with slender, grassy leaves to ~1.5 m. Even the lizardtail flowerheads look grass-like until fat berries swell up on their surface, revealing its closer relationship to the arum lily family, though many botanists believe it deserves a family of its own. Mature plants need constant moisture as well as protection from the hottest sun. It is not difficult to raise from fresh seed planted into warm soil though patience is required.

Gymnostachys anceps

Hanguana (Hanguanaceae)

A big swamp plant from far northern Australia, *H. malayana* (no common name in English) with its broad, attractively marbled leaves is one of the most unusual 'grasses' in this book. It reminds me of some bromeliads. Often found in dense stands covering large areas where no other plants can establish, the colonies reach up to 2 m high with branching green flowerstems held above, eventually forming purple berries. It will grow in waterlogged soils or shallow ponds, but is intolerant of cold and is unlikely to thrive much south of the tropics. Easily propagated by division or from stem cuttings.

Hanguana malayana

Helmholtzia (Philydraceae)

Stream lily (*H. glaberrima*) is more grass-like than its relative *Philydrum* until the spectacular and vivid masses of bright pink flowers appear. Reaching up to 1.5 m high in flower, this is a subtropical plant from shaded rainforest gullies and may be found growing densely along streams.

Helmholtzia glaberrima

Despite this, it needs reasonable drainage and unlike its close relative Woolly frogmouth (*Philydrum*) will not survive flooding for long periods. In deeper shade it flowers poorly or not at all. Selected forms can be propagated by division, and fresh seed germinates readily.

Hemarthria (Poaceae)

Mat grass (*H. uncinata*) is a low-growing and not particularly distinguished-looking grass, but it makes an excellent low-maintenance native lawn. An adaptable plant, Mat grass tolerates short-term flooding, moderate levels of salinity, and remains reasonably green even through drought. It is easily propagated by division but tends to run fairly slowly, so it may need to be divided repeatedly to spread it out evenly enough for mowing.

Hemarthria uncinata

Hypolaena (Restionaceae)

These slender-stemmed plants form upright, slow-spreading thickets to ~70 cm high on seasonally wet soils, with the most widespread species, green-stemmed *H. fastigiata*, found in the coastal areas of most Australian states. The Western Australian *H. exsulca* is a dustier grey-green, but otherwise similar. Both species are very tolerant of drought once established and are easily propagated by division.

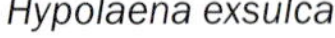

Hypolaena exsulca

Hypolaena fastigiata

Imperata (Poaceae)

Blady grass (*I. cylindrica*) is the native form of Japanese blood-grass, and although no native forms develop the rich red of that particular clone, in colder or drier conditions many of them will produce rich autumnal hues in shades of orange, gold and tan. The silvery, cylindrical, foxtail flowerheads are also attractive, but become untidy and should be cut off once they start to age and fall apart. A running grass to ~80 cm tall in moist soils, it is tolerant of drought, temporary flooding and poor, acid soils, and would probably make a good though rough lawn which would need only occasional mowing. Blady grass is easy to propagate by division, and it may be worth raising from seed in the hope of finding even more richly coloured variants to compete with the non-native Blood-grass cultivar.

Imperata cylindrica

Isachne (Poaceae)

Swamp millet (*I. globosa*) is a rather bamboo-like, running grass growing to ~50 cm high in seasonally dry shallows and more permanent waters. It will also grow in well-watered soil and makes a tidy specimen when kept in a pot. The small, delicate-looking massed seedheads create an attractive halo or haze above the foliage in late summer. It is easily grown from seed, and the individual plantlets that pop up along the runners can be separated and divided.

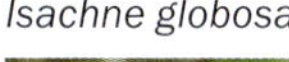

Isachne globosa

Ischaemum (Poaceae)

Large bluegrass (*I. australe*) is a coarse, running, subtropical grass from wet places among sand dunes, reaching close to 1 m high in ideal conditions, and with a vertically pronged flowerhead resembling some kind of spiky caterpillar, or in more compressed forms a head of wheat. Despite the common name, it is more likely to be reddish than blue in overall effect. Creeping wheatgrass (*I. triticeum*) is even more robust, with leaves more densely packed. Tolerant of salt spray, full sun and even short-term drought, these plants aren't difficult to propagate from runners, which take root easily when separated from the main plant.

Ischaemum australe in a coastal Wallum swamp, southern Queensland

Isolepis (Cyperaceae)

This group includes numerous species of relatively small, tufted sedges, mostly found around the fringes of wetlands and in damp places, and sometimes in shallow waters. Many of the smaller species are annual but a widespread species sometimes sold as 'Fairylights' (*I. cernua*) is much longer-lived. It can form handsome, weeping clumps tipped with tiny, pale flowerheads which make it look much like the once-fashionable fibre-optic lights which have rightly been relegated to the mists of time.

The foliage of Floating clubrush (*I. fluitans*) is also slender, forming floating carpets in flowing waters or dense lawns along streambanks and in marshes as water levels fall. The still coarser ribbons of Mat clubrush (*I. inundata*) trail elegantly in shallow streams, but if growing at the water's edge change completely and become upright tussocks topped by clusters of chocolate-brown spikelets. All species are readily propagated by division and adapt to a wide range of climates.

Isolepis inundata

Isolepis fluitans

Johnsonia (Hemerocallidaceae)

A small group of perhaps five Western Australian lily relatives, these look very much like some sedges and rushes when not in flower. The open, straight-stemmed tussocks of the largest species, Hooded lily (*J. lupulina*), reach ~60 cm and are particularly rush-like, though with the improbable addition of pale-pink pinecones towards the tips. The inconspicuous flowers are hidden inside the pink bracts of the cones, which may retain their colour through most of spring and well into summer. All species need good drainage and full sun in cooler climates, but will tolerate some shade in hotter areas, and are not difficult to raise from seed sown in spring.

Johnsonia lupulina

Joycea – see *Rytidosperma*

Juncus (Juncaceae)

The rushes are a large and diverse group varying from small, sprawling annuals to tall, upright tussocks in habit, though of the 60 or so native species only around six or seven of the leafless varieties are at all familiar to most gardeners. These are generally elegant and more-or-less upright, with their stems forming tidy tussocks, ideal even in a formal garden as long as you take the trouble to gently tug out dead, older growth once or

Juncus procerus

twice a year. Many species grow in seasonally damp places or even shallow waters, though a few such as the very upright, bottle-green Pale rush (*J. pallidus*) will grow in any normal garden soil. This species is named for its upright, pallid green flowerheads that age to a golden-brown, and it reaches ~1.5 m in height.

Some rushes only look their best if they have been flooded occasionally through the year, though even these may be fairly drought-tolerant once established. Giant rush (*J. ingens*) reaches 5 m in ideal conditions but is more usually just 2 m high, and has particularly large, intricate and pendulous golden-brown flowerheads. This species grows in abundant stands along sections of the Murray River and beyond to Sydney, and is a significant habitat plant for some waterbirds as well as being decorative. Its southern equivalent is Marsh rush (*J. procerus*) with more upright, deep-green stems and compact, deep-brown pom-pom flowerheads, reaching to 3 m and thriving even if flooded 50 cm deep through the winter months. There is even a widespread species which grows in tidal waters and saltmarshes around most of southern Australia, the aptly named Marine rush (*J. kraussii*) with blackish-green stems to 1.5 m, and glossy dark-brown flowerheads.

Another species which likes wet feet but will still look good in drier sites is Yellow rush (*J. flavidus*), a very upright plant resembling a slenderer version of Pale rush though reaching only two-thirds the height. Common rush (*J. usitatus*) may be the common species around Sydney but is increasingly replaced further south by the similar-looking Broom rush (*J. sarophus*), while Grey rush (*J. polyanthemus*) is a dustier blue-grey and more subtropical in range. All three will reach ~1.2 m or more and are upright plants with slender stems and brownish-red stem bases. In permanently moist, shaded situations they

Juncus holoschoenus

Juncus ingens female plant

Juncus ingens along the Murray River

Juncus pallidus

Juncus planifolius

will become softer-looking and more weeping in habit, turning a deeper green rather than the bluer, yellower or greyer tints of plants grown in full sunlight.

Some less familiar groups of rushes are very different from the leafless types described above, and while extremes such as the short-lived Toad rush (*J. bufonius*) are never likely to become popular because of their untidy habit, others have more potential in and around a water garden. Some with soft, hollow stems and masses of well-spaced reddish flowerheads such as Joint-leaf rush (*J. holoschoenus*) and its slender relative *J. fockei* are no

Juncus usitatus grown as a potted specimen

harder to grow than those already described, but they should never be allowed to dry out completely.

More surprisingly, some rushes are leafy and look remarkably like various blady grasses until you look closely at their flower structure (see Chapter 1). The best of these is Broadleaf rush (*J. planifolius*), which holds open masses of flowerheads 1 m above the compact mound of foliage. In the best forms of this species the base of each tussock turns a bright pink or red in strong sunlight, contrasting beautifully with the fresh green, grassy foliage above.

As many species of rush are potentially weedy if introduced to new areas and they adapt readily to a wide range of climates, it is essential to plant only species that are local (and ideally provenanced) to your area. This should be no hardship! There are a great number of species with a wide range of growth forms in all parts of southern Australia. All rushes are easily propagated by division (even sections of rhizome with no roots will usually establish readily), and their prolifically produced seed will germinate even on damp, nutrient-poor clay soils.

Kingia (Dasypogonaceae)

Drumstick grasstree (*K. australis*) remains one of the most undervalued, underplanted and least familiar of Australia's grasstrees. The more-or-less upright stems of old specimens may reach 8 m high, producing a crown or rosette of drumsticks after bushfires, rather than the upright spike of the more familiar grasstrees (*Xanthorrhoea*). These droop downwards as the seed matures and the growing crown moves on upwards. A south-western endemic, this remarkable and elegant plant remains reasonably common in the wild, and healthy relict specimens can even be found thriving around the fringes of pine forests!

Although regarded as a single well-defined species, plants from different areas may look markedly different, and though many of these are relatively broad-leaved my favourite form is the slender, silvery-leaved type from the Porongorups. The small populations north of Perth may prove to be more resistant to summer heat and drought than those from 700 km further south. Mature plants can be transplanted with care, though like other grasstrees they are dependent on nutrient reserves in the trunk. These may keep them looking reasonably healthy for years before they abruptly die – a disappointing waste of time and plants.

However, Drumstick grasstrees are not hard to raise from *fresh* seed and are faster-growing than most *Xanthorrhoea*.

Treat them like any other indigenous plant that is intolerant of unusually high quantities of nitrogen and phosphorus (the so-called 'native' fertilisers are usually acceptable) and, given that they are stimulated into flowering by fire, adding a little wood ash to their potting mix is likely to be beneficial. Some sources suggest that soaking the seed for a few hours in water with wood smoke dissolved in it will encourage germination, but it may take four months for the seed to

Kingia australis, a broad-leaved variant above, with drumstick flowerheads below

germinate in any case. Winters in the south-west can be both wet and cold for long periods of time so frost shouldn't be a problem for plants of appropriate provenance, but good drainage is essential to minimise the possibility of fungal problems.

Koeleria (Poaceae)

Crested hairgrass (*K. macrantha*) is a subalpine species closely related to *Deschampsia* and has also been suspected to be introduced, though the remarks under the latter species apply equally here. The tight, cylindrical flowerheads are silvery-white and carried on very upright stems to ~60 cm, but this species does not like hot summer climates so it has rarely been available in Australia. Propagation is usually from seed, but division is a more convenient option.

Lachnagrostis (Poaceae)

Blown grasses are efficient colonisers of exposed moist soils such as dry lake beds, but will also rapidly fill bare spaces among other grasses and shrubs in more heavily planted situations. In drought conditions in south-eastern Australia, Common blown grass (*L. filiformis*) may become the dominant plant over thousands of hectares of former wetlands.

Lachnagrostis filiformis

Although it attracts swans, coots and swamphens which feed upon it while it is lush and green, it can create a potential fire hazard when huge quantities of tumbling seedheads are heaped by wind against fences. Blown grasses form spectacular mounds of feathery plumes to ~60 cm in diameter. They are mostly annual, or at least very short-lived, but in compensation they reliably produce abundant crops of seeds for replanting.

Lepidosperma (Cyperaceae)

Swordsedges are a varied and adaptable group of useful habitat plants when massed along streambeds or even among coastal dunes and include many large and highly ornamental plants which make spectacular specimens in the garden. Even the most water-loving are well adapted to prolonged periods of serious drought, remaining fairly green and lush-looking though dead, older growth should be pulled out before summer to reduce potential fire hazard. They are by no means as flammable as sawsedges (*Gahnia*). The leathery, broad-leaved species with strap-like foliage are the only ones usually seen in cultivation, although there many attractive smaller types forming dense, slender-leaved tussocks with contrasting dark flowerheads. Some of these are not difficult to raise from seed.

Lepidosperma concavum

Lepidosperma gladiatus

Not surprisingly, the best-known species, including many of the most attractive, are widespread from Western Australia to the south-east. Of the more water-loving species, Tall swordsedge (*L. elatius*) grows along rainforest gullies which may only be seasonally wet but it will grow best in permanently moist soils. The wide, flattened leaves of these open tussocks reach 2 m long and are tipped with long, golden-brown, weeping flowerheads of particular elegance. Pithy swordsedge (*L. longitudinale*) is a similar height but more upright and slender. It grows in more

Lepidosperma longitudinale

exposed places which are seasonally flooded, yet may dry out for years at a time.

Another group of swordsedges is found in much drier places including sand dunes. Coastal swordsedge (*L. gladiatum*) is the largest at ~1 m high, with dense, cocoa-brown flowerheads ageing to silver-grey. Sandhill swordsedge (*L. concavum*) is a more compact version often not much more than half that height, with blackish flowerheads tipping leaves which may develop attractive autumnal colours in hot, dry weather or even in winter cold. Although both species are extremely drought-tolerant, they need to be watered occasionally during the first summer after planting, while their roots are still establishing. Swordsedges aren't always easy to grow from seed, which may need to be stored for a year or so before it will germinate. Division is more reliable, though they resent being transplanted and may take time to recover.

Lepironia articulata

Lepironia (Cyperaceae)

Spear-sedge (*L. articulata*) is one of the most dramatic native sedges, with stiffly upright, powdery blue-grey stems to 2 m high, with tiny, dark cones near the tips. With time it will form extensive colonies in ideal conditions, but it isn't clear whether it does this from seed or runners. Although this is an abundant wetland plant (standing in up to 1 m of water for months at a time) found from central New South Wales northwards to Malaysia, and provides a significant and drought-tolerant habitat type, it doesn't seem to have any widely accepted common or Aboriginal name.

It has been widely used in South-East Asia for various

types of fibre and mats including sails. It was possibly introduced to Madagascar by the first settlers from Indonesia many centuries ago for this purpose- or perhaps some seed just arrived there attached to old sails. The starchy roots have been cooked by indigenous peoples. Seed is not hard to germinate but as each plant only produces tiny amounts over a long season and the sharp stem tips are a danger to your eyes when collecting it, Spear-sedge is most conveniently propagated by division.

Leptaspis (Poaceae)

L. banksii is a bamboo-like grass from tropical Queensland, reaching ~60 cm high and growing in tufted clumps with the stems often concealed by the foliage. The flowerheads aren't anything to write home about, though they have the sedge-like habit of segregating male flowers at the top with females lower down. The leaves are long and fairly narrow with a prominent midvein and can reach 30 cm long. This grass was in cultivation in the early 2000s but it doesn't seem to be available from nurseries at present. It is not uncommon and can be grown even in subtropical areas if given protection from hot sun.

Leptocarpus canus

Leptocarpus (Restionaceae)

These handsome cordrushes grow in seasonally flooded sandy heathlands and along streams but will also grow on most well-drained soils if they have enough water through winter and spring. They tolerate a fair degree of drought once established. This genus has had a fairly complex history of name changes, with the former genus

Meeboldina being shifted here in 2014. Male plants tend to be weeping with masses of fine, pendulous, silvery flowerheads, very delicate in appearance, while female flowers are generally reddish and clustered into rounded or upright heads.

Slender twinerush (*L. tenax*) is the most widespread species. It is found in both eastern and south-western Australia, sometimes in places where its roots may be flooded for months. This species is moderately salt-tolerant and may reach up to 1 m in height. The stems of the south-western *L. laxus* reach twice that height and become still more pendulous under the weight of the flowerheads, males with their long cascades being particularly attractive.

Of the species formerly in *Meeboldina* (all of them from Western Australia) only a striking female form of *L. scariosus* with dense heads of red-brown flowers has been reliably available in eastern Australia, though the somewhat smaller *L. denmarkicus* is also available from some nurseries. *L. canus* is common around the Perth area and is easily grown, forming upright tufts with club-like female flowerheads or strings of pendulous male flowers. All these species reach ~70 cm high, some set seed freely and are readily raised on wet soils as described for *Chaetanthus*, but most are easily propagated by division in spring.

Leptocarpus laxus male

Leptocarpus scariosus female

Leptocarpus tenax female

Leptocarpus tenax male

Leptochloa (Poaceae)

Brown beetle-grass (*L. fusca*) is best known as a weed of rice fields and boggy places, but young plants with their leaden blue, weeping flowerheads on 60 cm stems, ripening to chocolate-brown, are distinctive and attractive. Growing as an annual in seasonally flooded places though often longer-lived in a well-watered water garden, they will self-seed and compete effectively with the many less-attractive introduced grasses which would otherwise take over under such conditions.

Leptochloa fusca

Libertia (Iridaceae)

Pretty grass-flag (*L. pulchella*) is an adaptable, open tussock reaching ~50 cm in height, found from Papua New Guinea to Tasmania and tolerating frost and a wide range of soils as long as it is kept moist. The starry white flowers appear from spring well into summer. Branching grass-flag (*L. paniculata*) with broad, rather iris-like leaves is similar but does not flower as long, and both are easy to maintain in pots. These plants are usually propagated by division: seed is reputedly slow to germinate.

Lomandra (Asparagaceae)

Whether you go to a native nursery or a garden centre, the dominant plants in the native grass section are likely to be mat-rushes with new varieties, selected forms and occasional hybrids appearing nearly every year. Their many good points include that most are low-maintenance and drought-tolerant, though they all look their best if given an occasional soaking during particularly hot weather. All named varieties

Lomandra 'Aussie Blue'

of mat-rushes are propagated by division, and many species and local forms are not difficult to grow from seed.

Though some of the larger species have striking and unusual flowerheads, the smaller and more compact forms mainly vie in how inconspicuous their flowerheads can be. In the wild they are mainly plants of woodlands, growing where they are shaded at least some of the time, and sometimes forming extensive stands on soils that retain moisture that bit longer. Even in drier areas they will be most abundant where rocky outcrops shade their roots and act as mulch to retain moisture for longer.

Some of the smaller and more slender-leaved green forms of *L. confertifolia* don't reach much above 30 cm in height, for example 'Little Con', 'Little Pal' and the hybrid 'Lime Tuff', reputed to tolerate even heavy frosts. Larger selections, reputedly of this species, such as the bluish-grey 'Seascape' are more upright and much taller – check their ultimate size before planting, or they may end up too crowded. There are numerous wildlings such as *L. cylindrica* which grow similarly and look equally interesting when in flower, particularly the lime-green 'Lime Wave' with slender, bright-green, semi-weeping foliage.

Lomandra confertifolia at left, with *Leptocarpus tenax*

Larger plants including numerous selections of the widespread Spiny-headed mat-rush (*L. longifolia*) make fine specimens, ranging from the relatively slender 'Tanika' with bright-green foliage, to robust strap-leaved plants with hefty flowerheads standing over 1.5 m high in shaded situations. Most cultivated selections of mat-rushes are of relatively widespread species, and these can vary dramatically depending on the local conditions. The most heat- and drought-adapted forms are often protected by a waxy bloom which gives them a distinctly bluish colour, and even the Spiny-headed mat-rush is now available in a bluish-grey form called 'Nyalla'.

Lomandra confertifolia subsp. *rubiginosa*

Lomandra cylindrica

Lomandra 'Aussie Blue'

Other less commonly available wild species are more pronouncedly blue, for example the elegant and very hardy Irongrass (*L. patens*) which reaches over 1 m tall, with masses of bronze flowers tucked away among the slender leaves. Some of the new garden forms may be hybrids, though

Lomandra longifolia flowering

Lomandra patens

this hardly matters from the gardener's point of view. Best known of these is the uninspiringly named 'Aussie Blue' which only reaches ~30 cm and seems to be a slow grower even in the strong sunlight it needs for best colour. *L. filiformis* 'Savannah Blue' is similar in size and perhaps faster growing, while *L. glauca* 'Blue Ridge' is a little larger and paler-leaved.

Loxocarya (Restionaceae)

The plants sold under this name in eastern Australia are correctly *Desmocladus*.

Macropidia (Haemodoraceae)

Black kangaroo paw (*M. fuliginosa*) is included here only because of its close relationship to *Anigozanthos*. Coming from warm, arid parts of Western Australia from Perth northwards, it grows on nutrient-poor laterite soils with plenty of iron but little in the way of other nutrients (phosphate is particularly deadly to it), and in humid climates it will pick up every fungal disease around. A plant strictly for people willing to put a lot of effort in for the sake of its lovely green and black flowers, rather than its grassy foliage.

Meeboldina (Restionaceae) – see *Leptocarpus*

Mesomelaena (Cyperaceae)

This group of Western Australian sedges includes some fascinating plants which stand out in almost any garden, despite their lack of showy flowers. Semaphore sedge (*M. tetragona*) is the most dramatic – a large,

Mesomelaena pseudostygia

Mesomelaena tetragona

open tussock to ~1 m high with long-spurred flowerheads like the horns of some primeval beast, growing on well-drained and relatively infertile soils by preference. The smaller tussocks of the widespread *M. pseudostygia* are only two-thirds that size, and are like a blackish-flowered version of Knobby clubrush (*Ficinea*) crossed with a sea-urchin. They prefer well-drained conditions, but *M. pseudostygia* will also tolerate fairly wet soils for months at a time. Rarely cultivated outside Western Australia, both species seem to resent disturbance of their roots, and are best propagated from seed.

Microlaena (Poaceae)

Weeping grass or Meadow ricegrass (*M. stipoides*) is widely regarded as probably the best native lawn grass for areas with reasonable rainfall, because it will form a tight, green, non-flowering carpet when regularly grazed by kangaroos or wallabies – this is known as a 'marsupial lawn'. Found throughout Australia other than in the drier regions, *M. stipoides* is one of the handful of native grasses which have been studied for their potential production of edible seed, though of course you can't maintain it as a lawn and expect a crop at the same time. Weeping grass is a variable species – it may grow as a small tussock and even includes some

Microlaena stipoides

furry-leaved forms. It is possible that different strains will eventually be selected for different purposes, from seed production to ornamental plants, as well as a range of appropriate indigenous strains best suited to use as a lawn in specific areas. Although it is easily propagated by division, larger quantities are usually grown from seed.

Molineria (Hypoxidaceae) see *Curculigo*

Muellerochloa (Poaceae)

M. moreheadiana is an unusual scrambling bamboo to 30 m or more in length from tropical Queensland, where it has occasionally been cultivated from root divisions. However, it is not likely to grow much further south and has rarely been available even from specialist bamboo nurseries.

Neurachne (Poaceae)

Foxtail mulga-grass (*N. alopecuroidea*) is one of the most distinctive, attractive, yet unfamiliar native grasses, with masses of tufted blue-grey flowerheads that almost seem to glow like some kind of smoky opal with the sun behind them. The tussocks form a low, spreading mound, with the flowerheads held well above on stems to 60 cm. Although this is mainly a plant of drier inland areas and prefers sandy soils in nature, it will also thrive in wetter areas in well-drained conditions. It is readily propagated from seed or by division.

Notodanthonia – see *Rytidosperma*

Oplismenus (Poaceae)

Beard-grasses are sprawling plants with relatively broad leaves spaced evenly along their elongated stems, and insignificant flowerheads at the stem tips. They are also known as basket grasses, for their habit of cascading up to 2 m from a suspended pot or basket. Basket grass (*O. imbecillis*, or possibly *O. hirtellus*) is widespread through the warmer parts of eastern Australia and in south-eastern Asia. The cultivar 'Variegatus' is most familiar, a handsome plant with broad, silver stripes on leaves flushed with pink on new growth.

The leaves of Rainforest beard-grass (*O. undulatifolius*) are similarly elegant though slightly crinkled and packed more densely on shorter stems.

Neurachne alopecuroidea

Oplismenus imbecillis 'Variegatus'

Creeping beard-grass (*O. aemulus*) has the broadest leaves, but as these are spaced widely it is the least ornamental of the three native species. All basket grasses need constant moisture, particularly when grown in a basket, but they will also grow as relatively low groundcovers to make a coarse and springy lawn in shaded places. Beard-grasses won't tolerate anything resembling frost and hate hot, dry sunlight, but are easy to propagate from divisions or cuttings.

Orthrosanthus (Iridacae)

Orthrosanthus multiflorus

The morning flags are a small group related to the purple-flags (*Patersonia*), with stiffly upright foliage. Their pale-blue flowers only open for a short time but may be freely produced over many weeks, with new flowerstems appearing from spring into early summer. The eastern Australian species *O. multiflorus* is most common around the Adelaide area though it is also scattered around Melbourne and across southern Western Australia, while *O. polystachyus* is most abundant in south-western Australia. Both reach up to ~60 cm high; the slightly larger Western morning flag (*O. laxus*) from Western Australia is also sometimes available. None of these species are fussy about soil (the plant photographed here was growing in pure sand on a coastal sand dune!) though they need full sunlight to flower well. Propagation is usually by division.

Oryza (Poaceae)

There are four native species of rice in northern Australia, including one that is believed to be the main ancestor of the world's most important cultivated grain. This is the annual Red rice (*O. rufipogon*), originally widespread through south-eastern Asia and probably first domesticated in southern China. A handsome plant with long-awned seedheads and orange grain, it grows to around 1 m in height and covers large areas of the Northern Territory floodplains as well as parts of northern Queensland. Unlike the domesticated forms of rice (*O. sativa*) which retain their seed for the convenience of harvesters, all wild species drop their seeds soon after they ripen. These are easily germinated in shallow, warm water on silty soil or mud, but are only likely to produce a reasonable crop in subtropical to tropical areas.

Oryza rufipogon

In recent years native populations of this species have been studied in some detail. It seems that these may have the greatest genetic diversity remaining in the world, making them potentially very useful for breeding new cultivars better suited to the accelerating effects of climate change. The diversity doesn't mean that domestic rice evolved from Australian stocks, rather that comparably diverse stocks in Asia have mostly been displaced by rice crops through most of their range. Ironically, most benefits from new rice varieties aren't likely to be particularly useful in Australia, as many of our waterbirds love rice. Magpie geese put an end to attempts at rice farming at Fogg Dam in the Northern Territory, descending on the growing crops in their tens of thousands!

Pandanus (Pandanaceae)

With ~30 native species, the screwpines are among the most distinctive small trees of tropical to subtropical landscapes. Although few have been tried in gardens, if you live near the sea Thatch screwpine (*P. tectorius*) of eastern Queensland and northern New South Wales (also widespread around many Pacific coasts and islands) is the nearest you will come to a salt- and wind-tolerant grasstree. Broad, saw-edged leaves spiral around the branches which carry large, scaly looking seedheads at the tips. The seedheads look a little like a stunted, woody pineapple and ripen to orange or sometimes red; they are edible but wild plants vary in quality and flavour. The tree reaches 5 m or more and is supported by thick, steeply angled prop roots around the base. Screw pines aren't hard to grow from seed partly buried in sandy soil, kept warm and reasonably well-watered for the first two or three years after planting, especially on sandy soils. Plants grown from cuttings in similar conditions establish and fruit much sooner.

Pandanus tectorius with Magnificent treefrog (*Litoria splendida*)

Panicum (Poaceae)

Australian Millet (*P. decompositum*, also known as Native panic) is closely related to so-called French or Proso millet (*P. milliaceum*) and was a significant bush food for Aboriginal peoples in many parts of mainland Australia, who ground it into flour. Widespread in drier inland areas, this drought-tolerant plant produces protein-rich, gluten-free seed that has attracted interest as a future potential crop. For gardeners interested in experimenting, however, it is still a long way from even the early stages of domestication and there don't seem to be any strains that retain their seed once it has ripened, so much of it is likely to fall before you have a chance to harvest it. And as one of our scruffier-looking native grasses, it is unlikely to have much of a future as an ornamental.

Patersonia (Iridaceae)

Native flags are distant relatives of irises, and 17 of the 19 species are Australian endemics. Many of them form wiry-leaved, upright tussocks and their three-petalled flowers are a particular attraction. The small and slender Short purple-flag (*P. fragilis*) of eastern Australia with its slender, grassy leaves through which the purple flowers peep is ~30 cm tall. The more iris-like Long purple-flag (*P. occidentalis*) comes in lilac shades (sometimes white), and its handsome, deep purple flowerstalks may reach 1.5 m. Long

Patersonia fragilis

Patersonia occidentalis

purple-flag is found in both eastern and western Australia, while the only yellow-flowered species Yellow flag (*P. umbrosa*) is a Western Australian specialty. Individual flowers only last a few hours but in some species new blooms appear for weeks. All native flags need full sun for best flowering, and most thrive in heathlands where they will tolerate wet feet for weeks or even months. All are easily raised from seed or by division.

Philydrum (Philydraceae)

Woolly frogmouth (*P. lanuginosum*) is a curious aquatic plant with thick, fleshy-looking leaves which are soft and spongy to the touch, and numerous small, yellow flowers appearing in succession over many weeks along the dozens of perfectly vertical flowering stems. In nutrient-rich soil these plants can form impressive specimens up to 2 m high in flower. Despite its soft and rather fragile appearance, this species will tolerate months of drying out, and such water-stressed plants develop a rich-red colour in those conditions as well as in colder weather. Easily propagated from the freely produced seed planted on waterlogged soil, or by rough division.

Philydrum lanuginosum

Phragmites (Poaceae)

Common reed (*P. australis*) is probably the only native grass almost all Australians recognise – though they may not be aware it is a grass – and worldwide it is one of the most familiar wetland plants other than cumbungi (*Typha*). The bamboo-like stems to 3 m tall are often found growing in fairly deep and permanent waters; in places that dry out over the summer months, plants will be shorter and slenderer. The plumed flowerheads can reach 50 cm long and look particularly attractive massed and moving in a high wind. Although it spreads by runners, Common reed can easily be constrained in a large pot shallowly submerged, or even miniaturised in a shallow laundry tub for the smallest water gardens.

In northern Australia, the even taller reed *P. vallatoria* (synonym *P. karka*) is found in more tidal waters or places close to the sea. Its flowerheads are larger and more open than those of Common reed in warmer climates, but both plants and flowers look scrappy and unimpressive in cooler conditions or purely fresh waters. Both species are salt-tolerant and are usually propagated by division, but Common reed is not difficult to grow from seed sown onto waterlogged, nutrient-rich soil if you need a few thousand plants in a hurry.

Although this brief account of the two species treats them as separate entities, the account of *Phragmites* in the *Flora of Australia* makes it clear that there is no simple, clear, defining characteristic that separates the two worldwide. In reality they are effectively a single, widespread and variable superspecies, even though in many places where they overlap there is no difficulty telling them apart. If the two were lumped into one species, for technical reasons it would probably become *P. vallatoria* and the far more familiar and widely used name *P. australis* would become obsolete. Fortunately, this is unlikely to happen as the resulting taxonomic chaos, affecting probably tens of thousands of research articles, wild populations and planting schemes in most countries, would arguably create the greatest botanical nomenclatural mess of all time! For some reason, I find this very amusing.

Phragmites australis

Phragmites vallatoria

Poa (Poaceae)

Tussock grasses are the definitive plants in a wide range of terrestrial habitats, from coastal fringes to high alpine areas, and include some of the most popular native grasses in cultivation. These plants can be very variable, and the more widely they range the more variable they are likely to be. In some cases it isn't even necessarily clear where one species ends and another begins, so don't rely on a single photo as a guide to the future appearance of seedling tussock grasses on offer at your local garden centre.

Used in mass plantings on roadside easements and traffic islands, tussock grasses are supposed to be low maintenance but on that scale they require brushcutting and regular removal of the abundant straw or they may become a fire risk. Indeed, sometimes they are set on fire to tidy them up – I have seen three lanes of a six-lane highway brought to a standstill due to billowing smoke from several hectares being burnt off on the broad median strip in the middle. With three fire engines in attendance and thick grey smoke adding to the greenhouse effect, this management strategy was neither low-maintenance nor ecologically sound. Also, fire may kill many of the tussocks through the brief but sometimes intense heat generated.

Poa costiniana

On a smaller and more judicious scale, mixed with other species of grasses to provide variety instead of an unimaginative, urban monoculture, the larger tussock grasses make fine specimens and add delicate textures if used sparingly along roads, so-called nature strips and along driveways. Common tussock grass (*P. labillardierei*) is by far the most widely used species in south-eastern Australia, and often forms magnificent flowering specimens to ~1.5 m high. This species thrives in permanently moist places though many of the best forms and selections look just as good with reasonable natural rainfall until after flowering has finished, changing to golden parchment shades as they dry out towards autumn.

The most common, adaptable and wide-ranging tussock grass in the wild, Common tussock grass is variable in colour. The bluish form collected for me by friends a quarter of a century ago from their Suggan Buggan property (in subalpine eastern Gippsland, Victoria) has been much admired even by Eurocentric gardeners over the years, but it is not particularly drought-tolerant or even all that blue when grown in warmer places than its subalpine home. The cultivar 'Eskdale' is more of a blue-grey, and holds its colour even in hot weather if kept moist.

Coastal tussock grass (*P. poiformis*) is more widespread, being found across south-western as well as south-eastern Australia. It is a generally lower and more open plant with relatively compact flowerheads, with

Poa labillardierei

some forms only 50 cm high. The finest forms I have seen were in Kings Park and the Botanic Gardens in Perth, a rich green with golden floral pokers. For gardeners who can't fit in the larger and sometimes more sprawling species, at two-thirds the size the sparser, grey-green Velvet tussock (*P. morrisii*) is an attractive alternative.

Blue- and grey-leaved species and forms mostly need full sunlight for best colour. Blue snowgrass (*P. sieberiana* var. *cyanophylla*) is an alpine form of a very widespread and variable species, with neat, low, waxy blue tussocks only a few centimetres high in cold conditions, holding numerous banners of jade-green flowerspikes up to 30 cm high. Despite its High Country origins, it will grow well at sea level in southern climates. The slightly larger Bog snowgrass (*P. costiniana*) is a little larger and greener, often with purplish stems in colder weather.

Fine-leaved snowgrass (*P. clivicola*) is an upright tussock to ~60 cm in flower, with greyish-green foliage which is particularly good in the selected form 'Mountain Blue'. There are many localised forms of *Poa* from higher and cooler climates that are worth keeping an eye out for in

Poa poiformis

Poa sieberiana

indigenous nurseries; for example, Horny snowgrass (*P. fawcettiae*) with open clusters of silvery-lilac flowers, or *P. ensiformis* from mountain forests of south-eastern Australia which forms tussocks comparable in size to Common tussock.

Many of these grasses are drought-tolerant in the sense that they survive adverse conditions, but don't expect them to remain green through a dry summer or in extremely cold winters. Instead, learn to appreciate their long-lasting dried flowerheads and sere, dormant foliage. Raking off dead leaf growth after autumn rains will reveal and encourage the new crop of green growth. All *Poa* species are easy to divide, and this is the only reliable way to increase particularly good colour or flowering forms, but most can also be raised without difficulty from their abundantly produced seed.

Pseudoraphis (Poaceae)

Spiny mudgrass or Moira grass (*P. spinescens*) is a running plant from inland areas of Australia including the Murray–Darling, and once formed extensive plains that have mostly disappeared due to a combination of grazing by livestock, damage by brumbies and ever-increasing and unsustainable water use for agriculture. Despite the common name, it

Pseudoraphis spinescens

grows happily on occasionally flooded sandflats and other places that never quite dry out, carpeting them with a blanket of lush green foliage and distinctive, wide open flowerheads in late summer. It is easily propagated from cuttings of the sprawling stems.

Rhynchospora (Cyperaceae)

Of the medium-sized sedges only *R. corymbosa* with its strappy, upright leaves and open flowerheads has been cultivated in Australia. Reaching ~1.5 m, it spreads gradually through shallow swamps and along slow-moving streams, looking like a somewhat ragged and sprawling leafrush. The more intriguing *R. rubra* has greater potential for a wider range of garden situations. Although reaching 1 m high in ideal conditions it is usually much smaller, with large, spherical flowerheads held high on separate stems above the foliage. This species appreciates the occasional flooding but will grow well in any constantly moist garden position. Both species love warmth and are frost-sensitive, and are usually propagated by division.

Richea (Ericaceae)

Like their close relatives *Dracophyllum*, this genus includes some fine-looking grasstree-like plants which are rarely available through nurseries. Nine of the 11 species are Tasmanian endemics, of which Pandani or Giant grasstrees (*R. pandanifolia*, named for their marked similarity to *Pandanus* screwpines including serrated leaf edges) are the most spectacular. These dramatic shrubs can reach over 10 m in height, forming a vertical columns or sometimes branching

Richea dracophylla

candelabra thatched with dead older leaves, presumably as protection from extremes of frost in their high-altitude habitats. The pink to white flowers are carried on branches off the main stem. Although fresh seed is not hard to germinate and cuttings may strike if you're patient, Giant grasstrees don't seem to do well outside their natural range, and aren't often seen in lowland gardens even in their home state.

The smaller *R. dracophylla* reaches up to 5 m high and is more of a branching shrub with leaves concentrated towards the end of the branches, where the columnar heads of creamy flowers also form. This species seems to be better adapted to relatively'warm conditions closer to sea level and has been grown as an unusual potted specimen, but it is unlikely to thrive in areas with hot, dry summers. Propagation is as for the Giant grasstree, though seed seems to be the more reliable source of new plants for this species.

Richea pandanifolia

Rytidosperma (Poaceae)

Wallaby grasses were originally grouped together as *Danthonia*, and were later separated into several genera: *Austrodanthonia*, *Notodanthonia* and *Joycea*. Now they are reunited, but in a new genus *Rytidosperma*. These tough, drought-tolerant grasses are often the dominant groundcover in dry, open eucalypt forest and in more open grasslands, but if you're trying to establish them in a garden they should be watered occasionally until their roots are well established. Wallaby grasses have been used as a rough lawn, but as they grow as separate tussocks the individual plants need to be closely packed if they are to look attractive once mown.

From a gardener's perspective many species look similar, being narrowly upright plants just a few centimetres across at the base, with silvery or creamy flowerheads held from ~70 cm to 1 m high. For this reason, local forms are likely to look just as attractive in the garden as any others from further away, so for new plantings stick to those produced by local indigenous nurseries. In south-eastern Australia the most widespread species include Long-leaved wallaby grass (*R. longifolium*) which may be close to 1 m in height, the slightly smaller but more densely packed Bristly wallaby grass (*R. setaceum*) found right across southern

Rytidosperma erianthum

Rytidosperma pallidum

Australia from east to west, and the relatively showy Hill wallaby grass (*R. erianthum*) extends even into arid inland areas.

Silvertop wallaby grass (*R. pallidum*) is perhaps the most distinctive species, a sprawling drought- and sun-tolerant blue-grey tussock to ~30 cm high which may be the dominant carpeting plant through extensive areas of dry, open bushland, and will often colonise disturbed ground even along road clearings. The more moisture-stressed it is, the greater the contrast between its silver-grey, older

Rytidosperma racemosum

leaves and the young green foliage partly hidden within. However, when too much dead material has accumulated it may be a good idea to tidy up by tugging off the dead growth in spring.

All wallaby grasses are easy to propagate from seed, though this may need a dormant period of several months before it will come to life. Selected forms are best propagated by division. Fertilising is not recommended as it is more likely to encourage the growth of weedier grasses, particularly introduced species. This is a common problem in re-establishing swards of native grassland.

Rytidosperma setaceum

Schizostachyum (Poaceae)

There is some doubt whether the unidentified bamboo known at present as *Schizostachyum* sp. 'Murray Island' is indigenous, as it has been suggested that it may be a relic of a plantation on Murray Island in Torres Strait. Given that this large-leaved and bushy clumper only reaches ~5 m and its canes are only 5 cm thick, it seems improbable that anyone would have imported and planted it when South-East Asia, only a few hundred

Schizostachyum sp. 'Murray Island'

kilometres away, offers hundreds of much larger bamboos with giant canes suitable for timber, edible shoots and many other uses. It is hard to see the logic in importing just one species of relatively minor usefulness, so Murray Island bamboo should be given the benefit of the doubt, especially as no one has managed to match it with a known species. Like most bamboos it is readily propagated by division, most easily done with smaller plants kept in pots until large enough to divide.

Schoenoplectus (Cyperaceae)

In ecological terms, these are the most important wetland clubrushes. They spread to form dense, erosion-resisting stands around the fringes, with the taller species growing into deeper, moving waters and providing nesting sites for diverse waterbirds (see also *Baumea*). These clubrushes will tolerate a certain amount of drying out, and most of them don't mind significant amounts of salt in their water. From a gardener's point of view, they are also among the most distinctive sedges with their diverse and

Schoenoplectus mucronatus

Schoenoplectus pungens

Schoenoplectus subulatus

attractive colourations, and leafless stems varying from cylindrical to sharply triangular in cross-section.

The largest species with cylindrical stems are usually found in deeper waters, with River clubrush (*S. tabernaemontani*) found from inland waters to within a stone's throw of the sea, though it will also form smaller plants in shallow or seasonal waters. For the record, this species has had several changes of scientific name, starting as *S. lacustris* and morphing into *S. validus* (this name is still in use for Western Australian plants) and most recently being classified as an indigenous form or variant of the northern hemisphere Common poolrush. In rich soils, River clubrush will reach up to 2 m high, and the blue-green stems with chocolate-brown cones near the tip make an attractive contrast to the superficially similar but bottle-green Jointed twigrush (*Baumea articulata*). However, gardeners would need a large pond to accommodate the two together!

Mangrove clubrush (*S. subulatus*) is even more salt-tolerant. It can be found in the upper reaches of tidal mangrove zones in northern Australia, though it also grows as far south as New South Wales. From the gardener's viewpoint it is perhaps not as attractive as River clubrush, as the stems are a less striking yellow-green and are not as tall. A distinctive

Schoenoplectus tabernaemontani

identification is the cylindrical stem that changes to a triangular cross-section near the flowerheads towards the tip. Two species with markedly triangular stems are widespread and readily available from water garden nurseries: Mucro clubrush (*S. mucronatus*, referring to a tiny leaflet near the base of each stem) and Delta clubrush (*S. pungens*). Mucro clubrush is a freshwater plant with bottle-green stems and densely packed heads of golden-brown cones near the fairly blunt tip, while Delta clubrush is a very salt-tolerant species found in inland saltmarshes and tidal river deltas. This is a more elongate species with waxy blue stems and less prominent flowerheads. Both species may reach 1 m in height in ideal conditions.

All species of clubrushes are easily grown in fresh water and will also thrive in varying degrees of salt or even brackish water. They are not fussy about soils, though they will be stunted if grown in clay, and are readily propagated by division or from seed.

Schoenus (Cyperaceae)

Many bogrushes are small annuals growing in seasonally wet soils, setting copious quantities of seed as the ground dries out. However, there are a few species with more promise in the garden. Creeping species such as Leafy bogrush (*S. maschalinus*) have potential as a dense-growing lawn that will never need mowing in permanently moist to wet places such as ditches, where it is rarely convenient to mow anyway! The most widespread and variable species is Common bogrush (*S. apogon*), although this is short-lived and may even be annual in drier places. The stems of this slender-leaved plant are topped with clusters of blackish flowers, forming a bushy plant usually no more than 20 cm high. Most perennial bogrushes form tightly woven tussocks of slender, more-or-less upright

Schoenus apogon

stems with tiny flowerheads, but Large-flowered bogrush (*S. grandiflorus*) from Western Australia has broader, strappy leaves and upright flowerstems to around 1 m, rather like a smaller version of some sawsedges (*Gahnia*). It is perfectly happy in most garden soils and conditions, and reasonably drought-tolerant. Annual bogrushes are propagated from seed, but most perennial species can be divided.

Scirpus (Cyperaceae)

Large-head clubrush (*S. polystachyos*) is a moderately tall and handsome subalpine sedge found along the higher, near-coastal ranges of south-eastern Australia. It will grow in permanent or seasonally flooded shallow waters, producing dramatic clustered flowerheads on 1.5 m vertical stems well above the foliage at the base around late spring. An unusual feature of this plant is a long, thick rootstock which resembles a woody carrot, though this is hardly a talking point unless you dig it up to show to visitors! Despite a wide natural range Large-head clubrush doesn't thrive in lowlands, especially where summers are long and hot. As seed isn't reliably set in cultivation, it must be propagated by division.

Scirpus polystachyos

Scrotochloa (Poaceae)

These somewhat bamboo-like grasses are strictly tropical and have rarely been cultivated, but would make attractive and unusual pot specimens even in cooler climates. Their culms are fairly solid and they have somewhat nondescript lower leaves, but leaves further up the stems reach 40 cm long and 15 cm wide and are subtly pleated. *S. tararaensis* is the larger and more interesting of the two native species.

Sorghum (Poaceae)

Grain sorghum (*S. bicolor*) is one of the five most important cereal crops in the world, so there is considerable interest in using its wild relatives (both annual and perennial) to create greater diversity and resilience to climate change in one of our major food sources. The 17 or more native sorghum species are reckoned by some to contain ~90% of the group's worldwide genetic diversity, so there is plenty of potential to use them to improve the cultivated crop in many ways, such as increasing drought tolerance and creating perennial cultivars that can be harvested year after year with minimal tillage or other soil disturbance.

At least some of the sorghum species were Aboriginal food plants, but as was generally the case early European invaders didn't bother keeping records so most indigenous knowledge is probably now lost. As the future breeding uses of native sorghums are still just hovering on the horizon, all I can report is that from a gardener's point of view few of these species have been planted as ornamentals. Nearly all of them are tropical or at best subtropical, and though many are drought-tolerant their relatively small tussocks are usually fairly sparse, especially in drier conditions.

Sowerbaea (Liliaceae/ Anthericaceae)

This small genus of purplish- or pinkish-flowered tussocks is mainly Western Australian. Vanilla plant or Rush lily (*S. juncea*) is the only eastern Australian species, and the only one regularly cultivated. The slender-leaved, upright tussocks grow to 50 cm and more, and are crowned with heads of small, starry, dangling, purplish flowers in spring. They will grow in any reasonably drained soil in a brightly lit place,

Sowerbaea juncea

and the most free-flowering forms are usually propagated by division though it is not difficult to raise them from seed.

Sparganium subglobosum

Sparganium (Sparganiaceae)

Floating burr-reed (*S. subglobosum*) is the only native species of this family. It is an unusual and fairly upright plant with slender, fleshy, triangular leaves, mainly growing in shallow fresh waters and not at all drought-tolerant. Reaching a maximum height of ~1 m and spreading by runners, it can be recognised immediately when in flower because of its mace-like balls staggered along somewhat zig-zag stems. Easily contained in a pot in a water garden, it can be propagated by division and from fresh seed.

Spinifex (Poaceae)

Coastal plants of this genus are very different from inland spinifex (*Triodia*, more appropriately known as Porcupine grass). These are the running grasses that hold sand dunes together, forming long chains of silvery tufts which are not only elegant but literally define the shape of many of our coasts. For many decades the western European Marram grass (*Ammophila arenaria*) was deliberately and misguidedly planted to control erosion around southern Australian coasts, presumably because Europhiles assumed it would do a better job than any native species.

The results of this misguided thinking are visible even from a distance. Marram creates a dune landscape more suited to the less exposed coastlines of the northern Atlantic, building much steeper and more vertical dunes than the native grasses which have adapted to our coastlines over untold millions of years. Marram dunes are more likely to blow out in the gales which build up along the Roaring Forties, where there isn't much land to slow the powerful winds from the west. By

Spinifex hirsutus seedheads

Spinifex sericeus, male flowerheads at right

contrast, the smoother and lower-lying dunes held together by coastal spinifex grasses offer less resistance to the worst storms, and thus hold their shape and integrity better.

In eastern Australia the main species is Hairy spinifex (*S. sericeus*, sometimes confused with the more western *S. hirsutus*), with seedheads that are relatively small though still impressive at up to 30 cm diameter. These resemble a giant pincushion, and are designed to tumble and blow over long distances before dropping their seed.

Along the warmer shores of the west the common species is Beach spinifex (*S. longifolius*), with even larger and more densely bristled seedheads, which paradoxically make them look softer from a distance. Coastal spinifex species can be propagated by division of the rooted runners, but these should never be taken from exposed situations as doing so will potentially trigger erosion. The seed is not hard to germinate but can be tricky to grow in containers as it is easily overwatered, especially in the cooler months. There is no reason to introduce either of these species outside their natural range for ornamental purposes; the coast has already been mismanaged without adding potential native weeds.

Taraxis (Restionaceae)

Slender-stemmed cordrush (*T. grossa*) can form dense thickets of 1.5 m stems through other plants it scrambles through, and is usually covered in swollen flowerspikes caused by galls. The cause of the galls is not known, so this plant should only be grown within its native range in south-western Australia, to avoid spreading potential pests or diseases. It does not seem to set seed even in the wild, but is easily propagated by division.

Taraxis grossa

Tetrarrhena (Poaceae)

Forest wiregrass (*T. juncea*) of south-eastern Australia is the most familiar species in this small group, a slender-stemmed, climbing grass which can form impressive mounds over treetrunks and fallen sticks, or scramble 8 m up taller plants. Despite their fragile appearance, the stems are tough and wiry with a rough and clinging finish, but it is easy to keep them tidy by lifting any stems heading off in the wrong direction and just draping or twining them where you want them to grow. In a bushland garden setting you may want to occasionally remove dead gum leaves caught among the stems so they don't build up too much, but this doesn't take long. The equally slender flowerheads of this plant aren't particularly striking, but in sunlight create a delicate-looking, shimmering mass over thickets. Wiregrass tolerates full sun in reasonably damp places but is even more vigorous in shade, and is easily propagated from root divisions – if you are willing to scrabble through the abrasive stems to find them!

A dense mat of *Tetrarrhena juncea* girdles a eucalyptus trunk

Thelionema (Asphodelaceae)

Tufted granite lilies (*T. caespitosa*, *T. grande* and *T. umbellatum*) are native herbs with long, narrow leaves around 1 m in length in ideal conditions and small, starry blue flowers from spring to mid-summer. A rather insipid white-flowered form of *T. caespitosa* called 'Montrose' is also available. Found naturally from central coastal New South Wales into Queensland at

Thelionema caespitosa

fairly high altitudes, these are adaptable plants that need reasonable drainage but will grow and flower in full sun or light shade, and tolerate light frosts. The species cross-pollinate readily but seed can be tricky to germinate, so new plants are usually propagated by division.

Themeda (Poaceae)

Kangaroo grass (*T. triandra*) is one of the most widespread southern hemisphere grasses, and was once a major component of grasslands throughout Australia. Sometimes referred to by the long-obsolete synonym *T. australis*, it extends through New Guinea to the warmer parts of Asia and across to southern Africa, where it is known as Red oat grass. This is one of the most widely used indigenous grasses in landscaping, with its impressive heads of rich-brown, oat-like spikes, and adaptability to diverse conditions including drought tolerance.

Most populations reach only around 1 m in flower and the tussocks themselves are only half that height, but I have seen spectacular local forms approaching 2 m tall in parts of southern Queensland, with massive heads that sweep dramatically in even the lightest winds. A relatively sprawling, low-growing selection called 'Mingo' is bluer than typical, and 'Big Blue' is over 1.5 m tall. Oat kangaroo grass (*T. avenacea*)

Themeda triandra

is an inland species from drier areas of the mainland and may reach more than 2 m. Both species can be propagated from seed, their named forms only by division.

Tremulina (Restionaceae)

Of the two species of Western Australian cordrushes, only *T. tremula* is available in eastern Australia. The densely jointed and flattened, rich-green stems form upright thickets to 60 cm in height on seasonally damp to wet ground. Both species can easily be propagated from seed which is freely produced, or by division while they are actively growing in spring.

Tremulina tremula

Triglochin (Juncaginaceae)

Streaked arrowgrass (*T. striata*) grows as a dense, low mat of upright leaves in shallow water or on mud, producing numerous attractive, upright flowerstems densely beaded with clusters of green flowers which retain their colour while the seed forms and ripens. The leaves are flattened in nutrient-rich situations, slender and cylindrical in drier conditions. Plants are easily propagated by division.

Triglochin striata

Triodia (Poaceae)

Porcupine grasses or desert spinifex (very different from coastal

Spinifex) are a large group of arid-country grasses found across most of inland Australia. They are handsome and soft-looking grasses but at the tip of each stiff leaf is a sharp spike – one touch is enough to deter anyone from experimenting with them again. Despite this formidable armature there are records of Aboriginal peoples harvesting the seeds of some species and grinding them to make nutrient-rich seed cakes, although these must have been only an occasional meal as seed production in desert conditions is unreliable.

Their slow-spreading, dusty-looking bluish to grey-green mounds may be up to around 1 m high in good conditions, and carry numerous stems of long, elegant, nodding, straw-coloured flowerheads. They make a striking display even in gardens which only receive a few centimetres of rain in an average year. Porcupine grasses adapt well to cultivation in wetter climates and will grow rapidly with the occasional watering in their first year or two, after which they can be left to their own devices.

In desert regions, sprawling old specimens can be found with their centres long dead and decayed away – see Chapter 3 for a discussion of how long native grasses can live. There are dozens of species, some still unnamed, in Western Australia alone, of which only *T. danthonoides*

Triodia irritans

seems to be cultivated. The previously widespread use of the name *T. irritans* to cover a number of species now recognised as different has caused some confusion, but identification can be simplified to some degree by geographic area; for example, the only species in Victoria is *T. scariosa*. The few species which have been cultivated are all raised from seed, as only a masochist would try to divide and repot older plants.

Typha (Typhaceae)

Cumbungis are the tall and upright, poker-headed wetland rushes also known to many people as bulrushes. There are two native species, both with rich-tan flowerheads, as opposed to the weedy, introduced and

Typha domingensis

increasingly common Cat-tail (*T. latifolia*) which has dark chocolate-brown pokers. Regardless of origin, these plants all run too freely for dams and larger ponds, though the smaller forms are not hard to keep contained in a pot. The pokers disintegrate into a mass of fluff as they mature, but if cut before this stage they can last for years as a dry flower.

Typha orientalis

Broadleaf cumbungi (*T. orientalis*) reaches 3 m tall so the largest forms of this species are of no use to gardeners, though there are many smaller forms which are less problematic. Narrowleaf cumbungi (*T. domingensis*) rarely reaches even 2 m and in many forms has extremely ornamental poker heads as narrow as 5 mm, up to 40 times as long as they are wide. The young shoots of both species are pleasant to eat, the abundant pollen from the male flowers (which form a short-lived poker above the longer-lasting female poker) can be used as a type of flour, and the roots can be crushed and strained to produce a high-quality edible starch. Seed is easy to raise but seedlings can be wildly variable, so the best ornamental forms are propagated by division.

Vallisneria (Hydrocharitaceae)

You can create a lush lawn even in a deep pond using eelgrasses, which can cover extensive areas in clear waters but also tolerate murky water or heavily shaded conditions. Growing in shallow silt and spreading by runners (from which they are easily propagated), their upright leaves trail up to the surface of the water and may vary in length to match the depth. Seed would probably be easy to germinate if it wasn't so rarely found, due to what must be one of the most inefficient pollination systems in the plant kingdom. Female plants flower at the emergent tips of specialised spiraling stems, while male flowers are released as tiny rafts that float to the surface then drift around randomly. As male and female plants are rarely found growing near each other (one native species is only known from female plants), fertilisation is an unlikely event.

There are a number of dark-green native species, of which Slender eelgrass (*V. nana*) with its bootstrap-width leaves is the most delicate looking. At the other extreme, Giant eelgrass (*V. australis*) may be 3 cm and more across the leaf blade. Both species are cold-tolerant, but the more tropical Cauline eelgrass (*V. caulescens*) is perhaps the most intriguing with a broad, paler midrib and foliage arranged along vertical stems, varying from lime-green to beetroot-crimson when grown in full sunlight. Erect eelgrass (*V. erecta*) is a distinctive species from far northern Queensland with unusually thick and fleshy upright leaves rather like those of some wateribbons (*Cycnogeton*), with the tidy habit of growing up to the water's surface and stopping there instead of trailing on indefinitely.

Vallisneria caulescens colour variations

Vallisneria erecta in an aquarium

Vetiveria (Poaceae)

Australian vetiveria (*V. filipes*) is a primarily tropical grass found on riverbanks and floodplains as far south as northern New South Wales. This is a fairly coarse-looking and long-lived species with pleasantly scented foliage forming mounds to ~1 m high, but it is not particularly ornamental in the garden. It is usually propagated by division.

Xanthorrhoea (Xanthorrhoeacae)

Grasstrees were all the rage among gardeners in the late 1990s. Mature plants were stripped from the wild, stuffed into pots and sold to garden centres, with the profits from this industry helping to partially offset the costs of unsustainable clearing and logging of native forests. Most of the grasstrees uprooted like this didn't survive long, as they gradually depleted the starch reserves they had accumulated over decades, or in some cases centuries. Their demise was hastened by gardeners who burnt off the leaves or singed the trunks to trigger flowering, which used up still more of the plant's ever-diminishing reserves.

There are ~30 grasstree species, all native to different parts of Australia. Few of them are commercially available, and many are protected because of their limited remaining range. They make handsome specimens with tall flowerspikes and tousled heads of slender leaves springing from elephant-foot trunks, and ancient multi-trunked specimens are particularly impressive. Even the young flowerheads are fascinating as they emerge, a deep, velvety green at first, turning to cream as huge numbers of starry, sweetly fragrant flowers open.

The common species of eastern Australia is Austral grasstree (*X. australis*) which can reach 3 m high, though it often starts to lean eccentrically (and attractively) before that stage. In south-western Australia, the Western grasstree or Balga (*X. preissii*) is the equivalent local species, and South Australia has the Yacca (*X. semiplana*). There are many much smaller species that may be better adapted to pot culture, of which the widespread, trunkless Small grasstree (*X. minor*) only reaches 50 cm or so, with slender flowerheads to ~1.2 m. My favourite of the relative miniatures is the Sand-heath grasstree (*X. caespitosa*) from western Victoria to southern South Australia, also apparently trunkless but forming multiple heads as the plants mature.

Immature flower spikes of *Xanthorrhoea preissii*

This species seems more willing to flower than most and has tan pokers carried on pink stems, but it is the foliage that particularly appeals to me – tidy, flaring rosettes of relatively wide, somewhat triangular leaves up to 7 mm wide.

Most grasstrees can easily be raised from their abundantly produced seed, and some nurseries have had interesting results from developing more distinctive and vigorous strains/species and increasingly

Xanthorrhoea caespitosa, young flowering spike

A *Xanthorrhoea glauca–johnstoni* hybrid

experimenting with hybrids. Although there seems to be little information on how it began, a particularly vigorous seedling of *X. johnsonii* seems to have been the basis of various earlier crosses (e.g. 'Supergrass') that are very fast-growing, developing a 30 cm trunk in just 10 years in some cases. The blue-grey leaves of *X. glauca* often appear in its hybrid progeny, which also show faster growth than the species itself.

All grasstrees need good drainage and are very sensitive to cinnamon fungus (*Phytophthora*) so if this is present in nearby bushland you are unlikely to succeed in growing them for long. Species grasstrees are generally not difficult to raise from seed but take 15 years or more to even begin to develop a trunk, so if in later life you want to proudly display a home-grown patch of your local species to other gardeners, you should start planting seed in your 20s or 30s.

Xyris (Xyridaceae)

Slow-growing yellow-eyes are mostly small, open tussocks varying from 40–60 cm high, growing in seasonally wet places, with their cone-like flowerheads held well above the foliage. The bright yellow or cream

Xyris operculata

flowers pop out between the scales of the cone several at a time, over a period of days. Some species such as the tropical *X. indica* are more like tiny irises than grass-like, and these are usually annuals. Species forming slender-leaved tussocks are generally much longer-lived, with Tall yellow-eye (*X. operculata*) and Slender yellow-eye (*X. gracilis*) not uncommon in cultivation. Both will thrive in permanently wet situations but are quite drought-tolerant once established, and are most readily raised from fresh seed planted onto wet, peaty soil.

Further reading and recommended sources

THE REFERENCES BELOW INCLUDE MANY other useful sources for information on native grasses and their relatives, often linked to websites. Gardeners or growers thinking of establishing an extensive area of native grasses, whether as a lawn or the more ecologically diverse option of grassland, will find useful information on preparation and planting of the site (including brief videos) at http://www.nativeseeds.com.au.

Among general references I would like to single out *Flora of Australia*, particularly volumes 44A, 44B and 44C which deal with grasses (now available online at https://profiles.ala.org.au/opus/foa). Combined with the *Australasian Virtual Herbarium* at https://avh.chah.org.au this has been an excellent way to keep track of name changes, particularly as the website directs you to the correct and current scientific names if you type in an obsolete name.

Briggs BG (2014) *Leptocarpus* (Restionaceae) enlarged to include *Meeboldina* and *Stenotalis*, with new Westen Australian species and subgenera. *Telopea* **16**(1), 19–41. doi:10.7751/telopea20147400

Chivers IH, Raulings KA (2015) *Australian Native Grasses: Key Species and Their Uses*. 4th edn. Native Seeds Pty Ltd, Melbourne.

Clarke I (2015) *Name Those Grasses: Identifying Grasses, Sedges and Rushes*. Royal Botanic Gardens, Melbourne.

Cowie ID, Short PS, Osterkamp Madsen M (2000) *Floodplain Flora: A Flora of the Coastal Floodplains of the Northern Territory, Australia*. Australian Biological Resources Study, Canberra.

Elliott M, Watsford P (2008) *Grasses of Subtropical Eastern Australia: An Introductory Guide to Common Grasses – Native and Introduced*. Nullum Publications, Murwillumbah, NSW.

Franklin DC (2009) Taxonomic interpretations of Australian native bamboos and their biogeographic implications. *Bamboo Bulletin* **11**(1), 25–29.

Galaihalage KSA, Myrans H, Norton SL, Gleadow R, Furtado A, *et al*. (2020) Wild sorghum as a promising resource for crop improvement. *Frontiers in Plant Science* **11** (online). doi:10.3389/fpls.2020.01108

Harden GJ (Ed) (1993) *Flora of New South Wales, Volume 4*. University of New South Wales Press, Sydney.

Henry RJ (2019) Australian wild rice populations: a key resource for global food security. *Frontiers in Plant Science* **10** (online). doi:10.3389/fpls.2019.01354

Kapitany A (2011) *Australian Gymea and Spear Lilies*. Kapitany Concepts, Melbourne.

Kapitany A, Marriott N (2017) *Australian Grasstrees:* Xanthorrhoea *and* Kingia. Kapitany Concepts, Melbourne.

McDonald S, Haslam S (2009) *Grasses: Native and Introduced Grasses of the Noosa Biosphere Reserve and Surrounding Areas*. Noosa Integrated Catchment Association, Tewantin, Qld.

McIntyre S, McIvor J, Heard K (Eds) (2002) *Managing and Conserving Grassy Woodlands*. CSIRO Publishing, Melbourne.

Marriott N, Marriott J (1998) *Grassland Plants of South-Eastern Australia*. Bloomings Books, Melbourne.

Meney KA, Pate JS (Eds) (1999) *Australian Rushes: Biology, Identification and Conservation of Restionaceae and Allied Families*. University of Western Australia Press, Perth.

Mitchell M (2002) *Native Grasses: Identification Handbook for Temperate Australia*. 3rd edn. CSIRO Publishing, Melbourne.

Romanowski N (2009) *Planting Wetlands and Dams: A Practical Guide to Wetland Design, Construction and Propagation*. 2nd edn. CSIRO Landlinks Press, Melbourne.

Walsh NG, Entwisle TJ (Eds) (1994) *Flora of Victoria Volume 2: Ferns and Allied Plants, Conifers and Monocotyledons*. Inkata Press, Melbourne.

Index

Page numbers for images are shown in *italic*.